E. H. WEBER:

THE SENSE OF TOUCH

E.H. WEBER (1795 - 1878)

E. H. WEBER:

THE SENSE OF TOUCH

De Tactu translated by H. E. Ross

Der Tastsinn translated by D. J. Murray

With a preface by J. D. Mollon and a historical introduction by the translators

1978
Published for
EXPERIMENTAL PSYCHOLOGY SOCIETY

By
ACADEMIC PRESS
London . New York . San Francisco

ACADEMIC PRESS INC, (LONDON) LTD.
24-28 Oval Road,
London NW1

United States Edition published by
ACADEMIC PRESS INC.
111 Fifth Avenue
New York, New York 10003

Library of Congress Catalog Card Number 77-92825
ISBN: 0-12-740550-X

Printed in Great Britain by
Whitstable Litho Ltd.
Whitstable, Kent

PREFACE

The number of experimental psychologists who read Latin is small, and is decreasing with unhappy rapidity. If translation of E.H. Weber's *De Tactu* had been postponed much longer, it is quite probable that it would have remained untranslated for ever. Even the great German texts are increasingly inaccessible to Anglo-Saxon psychologists and Weber's *Der Tastsinn* is much more often cited than it is read. Of those experimental psychologists who do remain fluent in ancient and modern languages, a significant fraction are graduates of the older universities. It is to an Oxford PPP graduate, Dr. Helen Ross, and a Cambridge graduate who took Part I in Modern Languages, Dr. David Murray, that we are indebted for the present translations. Dr. Ross is known for her work in sensory psychology and Dr. Murray has studied tactile memory; and they are thus peculiarly well equipped to prepare this definitive edition.

And it is a storehouse indeed that has been saved from distinguished obscurity. The prospective reader should not suppose that he will find in these pages only an account of some experiments on hefted weights and an exposition of the law with which Weber is eponymously coupled. Weber has interesting remarks to make on countless aspects of sensory psychology - on visual resolution, the binocular combination of colours, and the moon illusion, on summation, inhibition and adaptation in sensory systems, on the difference between simultaneous and successive presentation, and on selective attention, the externalisation of sensations and the difference between sensation and perception. Modern neuropsychologists have completely neglected Weber's systematically conducted and shrewdly discussed experiments on left-right asymmetry in the tactile and thermal senses. Weber's writings are not merely of interest to historians, for his observations retain freshness and, in many cases, an enduring validity. *De Tactu* and *Der Tastsinn* stand to the sense of touch as Helmholtz's *Handbuch der physiologischen Optik* and *Tonempfindungen* stand to vision and hearing; but whereas Helmholtz was a literature man, building upon the work of many predecessors, Weber was tackling, alone and brilliantly, an almost unturned field.

The present project reflects the continuing cooperation between the Experimental Psychology Society and Academic Press, but it is the first venture of its kind undertaken by the EPS. If it proves as valuable as we expect, the Society hopes to support further translations.

J.D. Mollon, Hon. Secretary, Experimental Psychology Society

This volume is dedicated to the memory of

Spike Tanner

CONTENTS

CONTENTS

INDICES

INTRODUCTION

In this introduction we shall give an outline of Weber's biography and then offer a short discussion placing the contents of Weber's writings in their historical context.

WEBER'S LIFE

Sources on the life of Ernst Heinrich Weber are few, and only a partial bibliography of his works is available (Weber, 1852; Dawson, 1928). However we can give a brief sketch based mainly on two anonymous obituary articles (1878, 1879), Ludwig's obituary speech (1878), Bück-Rich's monograph (1970), and various encyclopaedia articles.

Like Fechner, Weber was born into a family whose father was a Protestant minister. Michael Weber, himself the son of a peasant landowner, began as a preacher in Leipzig but was called to the University of Wittenberg as Professor of Dogmatics. He married the daughter of another minister, Lippold, and they had 13 children of whom 7 survived. Ernst Heinrich, the third child, was born on June 24, 1795; two other brothers, who themselves made excellent names as scientists, were Wilhelm Eduard, born in 1804, and Eduard Friedrich, born in 1806. Ernst apparently did well at school in Meissen, reaching the senior grade at the age of 12 and going to Wittenberg University at 16. He was proficient particularly at Latin, in which language he later wrote several monographs, and also at athletics. On entering University, he decided to take up medicine but even at this stage showed a particular interest in physics - one of the frequent callers at his father's house was the acoustician Chladni. However 2 years after Weber had begun his studies at Wittenberg, most of the lands belonging to the University fell into the hands of the Prussian army and many of the staff were forced to move to the small nearby town of Schmiedeberg. A short while later, the University of Wittenberg was obliged to close down and Weber's father took up a position at Halle; another cause of grief to the family was that Weber's mother died of consumption. Nevertheless Weber continued his studies and from Schmiedeberg moved to Leipzig where he devoted himself particularly to anatomical work under Rosenmüller. As his thesis (*Habilitationsschrift*) he wrote a work on the comparative anatomy of the sympathetic nerves; he himself illustrated the book. An important characteristic of his anatomical researches was his predilection for trying to link structure with function. This work was so well received that in 1820 he was offered a position at Bonn; however Leipzig offered him a similar position and he decided to stay there. It became his home for the rest of his life. We may note in passing that he appears to have been an excellent teacher (Bück-Rich, 1970).

His first research as a staff member consisted of an examination of the auditory system in fish: bones in this system are still known as Weber's ossicles. Later, in 1834, he extended his studies of audition to include a discussion of the role of the human cochlea in detecting sounds. Following Rosenmüller's death, he was appointed as *ordentlicher* Professor in Anatomy in 1821; he also married the sister of one of the friends of his youth. It was shortly after this that he began experimenting in physics as well as in biology. Noticing that particles lying on the surface of a phial of mercury showed certain patterns when the phial was moved, he invited his brother Wilhelm to participate with him on studies of wave motion in fluids. Some of their findings were based on experiments carried out in their father's house at Halle; Weber walked there from Leipzig every week, a distance of about 20 miles. Other findings were based on observations made in Alpine lakes, which the two Webers visited accompanied by Ernst's wife. The results were published as a joint monograph in 1825, and this work is still a classic of hydrodynamics. However Weber continued his physiological research, which included studies of the physical properties of hair and experiments on the vertebral column. The latter involved an investigation of the properties of the spine both at rest and in motion. With Wilhelm, he also wrote a treatise on the theory of musical instruments (1833), but the main work in which he combined both his knowledge of physics and biology appeared later in 1850: it concerned the application of wave theory to the study of blood flow. In this work, not only did he discuss the elastic properties of the arteries and veins, including measurements of the rate of propagation of the pulse-beat, but also made mention of various glands including the lymph vessels. More comprehensive works of the early period at Leipzig included a handbook of human anatomy, which was later to influence Müller's Handbook; and editions of manuals of anatomy by Hildebrandt and by Rosenmüller. Although from this account one might have the impression that Weber was a single-minded person devoted only to science, such was far from the case; he was elected to a senior administrative position at Leipzig and was able to use his influence to persuade the Saxon authorities to give a large grant for medical expansion at the University. Apparently his upbringing in a religious family helped him here, for one of the reasons he was successful in this endeavour was his argument that in the expanding industrial state of Saxony there was a strong need for religious education. More details of Weber's activity in academic politics will be found in the 1879 obituary. In 1840, following the death of Kühn, he also became Professor of Physiology at Leipzig. Other academic honours included membership of the Royal Saxon Society for scientific studies, and later, foreign membership of the Royal Society (1862).

Weber's best known work, however, dates from the period of approximately 1830-1850. Possibly as a result of his earlier work on the nervous system - which included pioneering studies of the nervous control of heartbeat and in particular the inhibitory properties of the vagus nerve - he began to study the nervous system of the skin and of the special senses in man. It has already been stated that Weber always asked how structure reflected function: his question now was, how does the structure of the nervous system lead to a functional representation of external space? One of his first findings - which has certainly been neglected by psychologists - was that the skin is perceived by the subject to have surfaces of varying extents and that the resulting map of external space is biassed accordingly. He arrived at this conclusion *via* studies of what we now call the two-point threshold. He also examined the perception of weight and temperature, and tentatively put forward what is now known as Weber's law. The results of this research, which were in part carried out with the aid of his brothers, appeared in *De Tactu* (1834). Ten years of further research followed, which led to a more precise formulation and expansion of his law and to a more comprehensive theoretical picture of sensation which included "common sensibility". This term is defined in more detail below. The later conclusions were summarized in an article written for Wagner's Handbook entitled "The sense of touch and common sensibility" (1846). (German: *Der Tastsinn und das Gemeingefühl*). The two monographs *De Tactu* and *Der Tastsinn* are translated in the present volume. It is worth observing that Fechner, who would elevate Weber's law to a primary principle in his new science of psychophysics, was at Leipzig during these years, although it is not clear just how much direct contact he had with Weber on these matters.

After these two masterpieces, Weber continued to work on other aspects of physics and biology. In biology he published a study of the male sex organ (1846) and discovered rudiments of the uterus in the males of various species. In physics, he examined the processes of fermentation with special reference to movements of particles in fluid (1854). His later life, according to Ludwig, was peaceful and settled: he had a harmonious family and a wide circle of friends. He continued to be an active athlete, being particularly good at swimming and skiing; and among his scientific contributions was a description of what is now known as Weber's syndrome, a form of hemiplegia involving a paralysis of the oculomotor nerve on the same side as the hemiplegia. In 1866 he gave up the chair of Physiology and in 1871 the chair of Anatomy. Weber died on January 26, 1878. His brother Wilhelm by then had become an eminent physicist after whom a unit of magnetic flux, the 'weber', is still named; and Eduard had become a physiologist with other administrative duties at Leipzig. Nearly all German biographical indices refer to all three and it is perhaps strange that, in view of E.H. Weber's importance in the history of psychology and of sensory physiology, he did not attract the same attention by historians of science as did Wilhelm. A more complete account in English of Weber's work on haemodynamics and on vagus inhibition will be found in Dawson (1928).

THE CONTENTS OF *DE TACTU* AND *DER TASTSINN*

Weber described nearly all his research on touch in the two books here translated. If one had to summarize the contents of these works as briefly as possible, it would be to the effect that *De Tactu* is heavily experimental while *Der Tastsinn* is broader and more theoretical. Many of the experiments reported in detail in *De Tactu* are described more briefly in *Der Tastsinn*, so there is some overlap, but the translators felt that the overlap was not so great as to warrant an abbreviated version of either work. In fact, of the 7 major tables in *De Tactu*, only 2 are repeated in *Der Tastsinn*. This section of the Introduction will therefore focus on an exposition of the contents of the two works, with emphasis on the relationships between them. We shall conclude with a brief evaluation of Weber's contributions.

De Tactu opens with the direct asking of the question: Does sensitivity to touch vary between the different bodily regions? In order to answer this, Weber first concentrates on the question of sensory acuity. He devised the now standard two-point threshold task in order to obtain experimental data on the subject: so far as we know, he had no predecessors in this respect. He carefully describes the two-point threshold for various bodily regions, showing, for example, that the oral region is more sensitive than the trunk to this kind of stimulation. He makes two additional observations of considerable interest: first, if the two points of the compass are laid transversely across the axis of a limb, sensitivity appears to be greater than is the case for the two points being laid longitudinally along the limb. His second novel finding is stated in Proposition VIII of the early part of *De Tactu*: we can better discriminate between two sensations falling on closely contiguous body parts if the two sensations are successive as opposed to simultaneous. Modern terminology would re-phrase this finding as: the two-point threshold exceeds the point-localization threshold. It has been amply confirmed since Weber's time (see notably Weinstein, 1968). But Weber is not content with a simple empirical description: he goes on to show that these findings are probably not artifacts of vision, the mechanical sense, or even the distribution of papillae: he suggests that the more dense the nerve-endings in the skin (what we should now call the 'density of cutaneous receptors'), the greater the sensitivity. From the outset Weber is looking for a physiological explanation of his results: this theme permeates both books.

But sensitivity to tactile stimulation depends on more than just passive reception of stimuli by the skin. We can obtain a better idea of the shape and texture of objects by actually moving our tactile surfaces over them. In so doing we obtain information both from the varying display of tactile sensations and also from sensations from the muscles (and more particularly, we now know, receptors in the areas of the joints). The estimation of weight, is better, for instance, if we lift an object than if the object is placed on a passive body-area. Weber discusses various aspects of this topic: it is of some historical interest that, although Brown (1820) and others had suggested there did exist a 'muscle sense', Weber was among the first to give it much attention in a physiological context. This point has been stressed by recent historians of psychology, particularly Murphy and Kovach (1972). Weber also devotes an interesting section to a fact which has been largely ignored in the recent past: an object of a certain weight is perceived as pressing more heavily on the left hand than on the right. We still do not have an adequate explanation of this phenomenon, though Weber suggests that the nerves on the left side are more 'sensitive' than nerves on the right side. In connection with this observation Weber points out, in a later passage, that temperature sensitivity is also greater in the left than the right hand. This, however, he ascribes to different thicknesses of epidermis in the two hands.

His discussion of the temperature sense, while not adding very much to our knowledge of the skin receptors mediating temperature, is of great historical importance because it broaches the problems of neural summation and adaptation. Again, he seems to have been the first person to experiment on the topic. As stimuli, he used either water of a certain temperature into which a body area was dipped, or (more particularly in *Der Tastsinn*) by touching the body area with a hot or cold object. In modern times von Békésy (1959) has shown the value of comparing the skin sense with the auditory sense; Weber, on the other hand, compares the skin sense with the visual sense. The fusion of temperatures is compared with the fusion of colours.

The last third or so of *De Tactu* is an attempt to draw theoretical conclusions from the experiments reported in the earlier part of the book. Weber stresses certain phenomenological features of touch sensation which physiologists might be tempted to overlook: most notably, if the two points of a compass are *moved* over the skin, they seem to 'spread apart' or 'come together' depending on the sensitivity of the area in question. This later became known as 'Weber's illusion'; it was confirmed in detail for 24 out of 42 different bodily regions by Goudge (1918). There is a strong hint of the 'sensory circle' theory later to be expounded in *Der Tastsinn*; there is the first mention of Weber's law, again to be expounded more fully in *Der Tastsinn*; and he suggests that heat and cold are not aspects of a continuum in the way light and dark are, but seem to be separate modalities of experience. Modern work on 'warm spots' and 'cold spots', while not completely confirming the hypothesis of separate receptor systems for warm and cold sensations, is nevertheless foreshadowed in Weber's discussion (see Geldard, 1972, for a review of this problem). It would not be too extreme to say that Weber's ideas on the touch-sense were almost complete by the time of *De Tactu; Der Tastsinn*, published 11 years later, adds more, to be sure, but it is at the level of elaboration rather than of innovation with respect to the major concepts of *De Tactu*.

After the pure experimentalism of *De Tactu*, the opening of *Der Tastsinn* comes as something of a surprise. Instead of going directly to the topic, Weber spends several pages discussing two problems which nowadays go under the names of 'the mind - body problem' and the 'sensation - perception' distinction. A few comments on this unexpected turn of events are perhaps appropriate. With respect to the former, we should recall that the 'mind - body problem' had preoccupied philosophers and psychologists, not merely from the days of Descartes, but from the very origins of psychology in Aristotle. By Weber's time theories on the matter ranged from Descartes' interactionism to Leibniz's parallelism. But the prevalent philosophy in Germany in the early 1800's was 'Nature Philosophy', most succinctly expounded in the work of Fichte and Schelling, but having its roots in Kant. Roughly speaking, nature-philosophy stresses the unity rather than the variety of natural phenomena. To give an extreme example, quoted by Bück-Rich, people saw links between the circulation of the blood and the fact that the earth rotates. Herbart, the most influential psychologist of the period, and the originator of the term 'threshold', discussed the matter extensively (Herbart, 1823): and nature-philosophy clearly influenced both Fechner and the writers of the Romantic Movement in Germany. Weber's achievement was to show that mental phenomena such as memories and images probably had their roots in physiological phenomena, a fact which tended to support the notion of a universe characterised by uniform laws. This was no idle speculation: the first pages of *Der Tastsinn* are full of examples of the close interlinkage between mental phenomena and bodily phenomena. They range from discussions of the moon illusion to Bessel's 'personal equation' (the origin of reaction time studies, see Boring, 1957). Weber's main focus was the question 'How do we form the concept of an external object?' His answer, as mentioned in the previous section, is that we do so by the way of interpretations of a mosaic of sensations. This of course brings us directly to the second problem, the distinction between 'sensation' and 'perception'. Most histories of psychology impute the distinction to Reid (1785) and then say little about it until they come to discuss Helmholtz' work on vision. But it is clear that Weber constitutes an important intermediate link between Reid and Helmholtz. The point we wish to make here is that Weber should be remembered not only for his experiments on the touch-sense but also because he contributed to the growing communication between philosophers, physiologists and psychologists in the early 19th century. It should also be noted that Weber follows Kant in believing that the concepts of space and time are probably innate and are superimposed, as it were, on concepts and interpretations arising from raw sensations. More complete discussions of the relation between Weber's physiology and the philosophy of his time will be found in the article by Bück-Rich and in the chapter on Weber by Kantor (1969).

Weber then turns to a consideration of the nature of the peripheral endings of nerves. He mentions Müller's doctrine of specific nerve energies, which had appeared in its fullest form 8 years before *Der Tastsinn*; and he provides interesting experimental evidence that it is the ending of the nerve, rather than the nerve-trunk itself, which appears to mediate sensation. It is perhaps worth mentioning at this point that, in Weber's time, the only cutaneous receptor ending known to physiologists was the Pacinian corpuscle, discussed in more detail in a later section of *Der Tastsinn*. The endings of Merkel, Krause and so on were not discovered until the latter part of the century. However Weber points out that conduction in the nerve-trunk can be prevented by pinching the nerve tightly. He goes on to discuss the rest of the somatosensory system but says relatively little about the pathways in the spinal cord. Instead he stresses that sensation is mediated not only by the nature of the peripheral stimulation but also by the way the fibres converge in the brain. Later he will adduce evidence from cases of brain-injury (mainly those causing hemiplegias) which supports this view. Weber, again, was a pioneer with respect to the use of clinical material to substantiate more purely theoretical speculations.

The next major section of *Der Tastsinn* is "on the sense of touch in particular". He points out that there are not merely separate sensations of pressure and temperature, but also that the two, under certain circumstances, become integrated. He reports the *Thaler* experiment which seems to indicate that cold objects feel heavier than warm objects. So far as we have been able to discover, this observation has not been confirmed in any substantial way in the literature since then, but L. Heiss, of the University of Michigan, did an undergraduate research project in 1969 on the matter. He found that cold objects did indeed feel heavier than warm objects and suggested that the reason might be that 'cold spots' are more common than 'warm spots' in most parts of the skin (see Hensel, 1973, for evidence of this). It could be that 'heaviness' judgments are in part determined by the density of receptors stimulated. Following some more general remarks on the somatosensory system, Weber then reports his earlier data on the two-point threshold - this is one of the points of overlap between *De Tactu* and *Der Tastsinn*. Again, the importance of the muscle sense is brought out.

His subsequent discussion of the temperature sense is an elaboration of that in *De Tactu*, but now Weber is more specific about the variables which influence judgments of temperature. He talks of the importance of heat conduction in the skin; of the thickness of the epidermis; of the size of the area stimulated; of the duration of stimulation; and of variations in temperature sensitivity in various body regions. The final portion of the section on cutaneous touch deals with broader problems, such as the question of whether simultaneous sensations tend to fuse or not. It is here that we get the fullest statement, in Weber's own writings, of Weber's law. He shows that the same law seems to hold for weight discrimination, visual discrimination, and tone discrimination. He never states the law in the mathematical form to which we are now accustomed, i.e.

$$\frac{\Delta I}{I} = k$$

where I is the 'intensity' of one stimulus and ΔI is the extra intensity needed for another stimulus to be just noticeably different from the first stimulus. It was Fechner (1860) who first put it in this form and who, of course, by the addition of certain assumptions, was led therefrom to Fechner's law. Speaking more generally, Weber's law has been confirmed many times since 1846, although it seems to break down at the extremes of the range of stimulus intensities in question in any one experimental paradigm. Perhaps the most interesting recent contribution to this literature is an attempt by Laming (1973, Chp. 7) to explain *why* Weber's law should take the form it does. It is too complicated to summarize here but depends on the ideal observer model of signal detection theory. On this matter, Tanner (1961) should also be consulted.

The last major section of *Der Tastsinn* is concerned with 'common sensibility'. This is a term which actually goes back to Aristotle, but which has completely dropped out of our scientific vocabulary since Weber's time. It seems to subsume all sensory experiences which are not obviously due to pressure or temperature changes. For example, in this section, Weber deals with pain, tickle, muscular fatigue, hunger, thirst and other vague sensations arising from vascular or secretory activity in the body. The most important point he makes is that even these sensations depend on activity in the nervous system. He shows that extremes of temperature lead to painful sensations (here, again, there is some overlap, with *De Tactu*) and raises the yet unsolved question of the nature of sensory receptors in the internal organs (on this topic see also Boring, 1915). He grapples indirectly with the question of 'double pain' (that is, a sudden 'pricking' pain arising from e.g. a pinprick, seems to be followed by a longer and more 'burning' pain). The modern theory is that this experience arises because fast-conducting nerve-fibres mediate the first sensation, while different nerve-fibres, which conduct more slowly, mediate the second sensation. This is perhaps the best example in Weber's writing of his ingenious attempts to explain unusual phenomena. It would be interesting to annotate all of this final section in the light of modern knowledge: there would probably be more gaps than solutions in such an exercise.

Weber's most innovative contribution to neurological theory was his concept of 'sensory circles'. They are referred to in passing both in *De Tactu* and in *Der Tastsinn*. The notion is that the skin is supplied by single nerves in such a way that if two sensory circles are simultaneously stimulated, two distinct sensations can be felt, but if only one sensory circle is stimulated simultaneously in different parts, only one sensation results. Murphy and Kovach state that the theory has lost ground because of the discovery of punctate sensation by Blix (1884) and others since: on the other hand Büch-Rich argues that the theory has received some confirmation from electrophysiological studies. But, whatever the modern opinion, the general idea that a single nerve innervates a larger region of skin than a mere point anticipates the discoveries of such broader phenomena as the dermatomes (Lewis, 1942) or the somatosensory areas of the cortex. One of Weber's most interesting ideas is that a 'sensory circle' might arise not simply because a nerve branches to innervate a given area but because it might be that a single fibre, receptive to stimulation throughout its end-region, snakes about under the relevant area of the skin surface. Another theme running through Weber's writing is his interest in electricity. We must recall that, at his time, it was known how to produce electricity from a Voltaic pile; certain electrochemical phenomena had been observed; the science of electromagnetism was in its infancy; and it had not yet been shown conclusively that a 'nervous impulse' was itself a propagation of electrochemical changes down a nerve-fibre. Nevertheless, as a physicist, Weber recognised the potential theoretical importance of electricity. It might also be remarked at this point that Weber's background in physics shows in other respects - his emphasis on competing forces when one exerts pressure on something, his emphasis on the heat-conducting capacity of bodily regions, his understanding of sound-transmission through the bones of the middle ear. As mentioned earlier, his main contribution as a biophysicist, however, consisted of his treatment of haemodynamics.

We may summarize Weber's original contributions as follows. He devised various experimental techniques for measuring the phenomena of sensation, notably the two-point threshold technique and the technique of studying temperature-sensations with the use of water of varying temperatures. He discovered the broad outlines of differential bodily sensibilities with respect to discrimination, localization, temperature, and (to a lesser extent) pain. At a more theoretical level, he stressed the interaction of brain with peripheral processes in determining sensory qualities; he established Weber's law; his notion of 'sensory circles' anticipated later discoveries; and his was among the first research concerned with integral phenomena of the nervous system such as summation, inhibition, fusion, and adaptation. More generally he must be recognized as a first-rate experimenter: he played an important historical role in the shift from philosophy to physiology as being the main contender for a psychologist's interest: and he made ample use of clinical material to establish certain ideas. It is our impression that many of these items have not been given a sufficient attention in the literature on the history of psychology and it is hoped that the present work may do something to remedy this state of affairs. Nevertheless Weber's books should also be read in their historical context. For example, here are some of the things about the somatosensory system we now know which Weber did not know. He did not know about most of the receptor endings in the skin: he knew nothing of the various kinds of electrical potential in the nervous system: he makes little mention of the spinal cord (though he did know of the distinction between sensory and motor nerves). He did not know how the cortex functioned: he knew little about nervous integration and reflex phenomena: and to some extent he was led astray by the concept of 'common sensibility'. He nowhere seems to suggest that pain is mediated by specialized receptors and/or nerve-fibres; and there is scarcely a mention of biochemistry (other than occasional reference to poisoning or to anaesthesia), although his section on muscular fatigue in *Der Tastsinn* is highly suggestive of a biochemical explanation of certain bodily sensations. Probably many readers will turn to the present work in order to read what Weber himself said about e.g. the two-point threshold or Weber's law; but it can also be seen as a document recording the battles and triumphs of a brilliant mind beginning research in an almost completely new area.

Finally, a word about the translation. Weber's style, both in Latin and German, tends to long sentences which sound cumbersome if translated literally into English. We have tried to avoid such awkwardnesses but make no promise that we have completely succeeded in doing so. He also uses many outdated units of measurement such as Paris lines, thumbs, joachims, and commas. These are explained in translator's footnotes at the appropriate places in the text. There are two kinds of footnotes: translator's footnotes, indicated by asterisks, and Weber's own footnotes, which are numbered consecutively through the text. The latter will be found at the end of each text. The page numbers of the original texts are given in square brackets at the nearest convenient break in the translation.

ACKNOWLEDGEMENTS

It only remains to thank the several persons who have contributed in the way of criticism, tracing of materials, and correction to the final translation. Both translators are extremely grateful to J.D. Mollon for the sensitivity and acuity with which he commented on earlier drafts of both *De Tactu* and *Der Tastsinn*, and for his encouragement throughout the undertaking. Helen Ross would also like to thank R. Campbell, G. Fisher, D. Gerver, F. Kiefer, W. Meischner, and P. Smith for various forms of assistance; and David Murray likewise expresses his appreciation to U. Ammermann, U. Frith, R. Macgregor, E. Scheerer, and E. Weinstock. Finally we must both extend warm thanks to the Committee of the Experimental Psychology Society and to Academic Press for their support and co-operation in producing the work; and to Mrs. K. Dorey for typing the final manuscript, and to Mrs. F. Lennie for preparing the index.

Helen E. Ross
University of Stirling
Scotland

David J. Murray
Queen's University
Kingston
Canada

REFERENCES

ANONYMOUS. Obituary of E.H. Weber. *Deutsche Zeitschrift für Praktische Medicin*, 1878, *14*, 165.

ANONYMOUS. Obituary of E.H. Weber. *Proceedings of the Royal Society*, 1879, *29*, XXVIII.

BÉKÉSY, G. VON. Synchronism of neural discharges and their demultiplication in pitch perception on the skin and in hearing. *Journal of the Acoustical Society of America*, 1959, *31*, 338.

BLIX, M. Experimentelle Beiträge zur Lösung der Frage über die specifische Energie der Hautnerven. *Zeitschrift für Biologie*, 1884, *20*, 141.

BORING, E.G. The sensations of the alimentary canal. *American Journal of Psychology*, 1915, 1.

BORING, E.G. *A History of Experimental Psychology.* Second edition. New York: Appleton - Century - Crofts, Inc., 1957.

BROWN, T. *Lectures on the Philosophy of the Human Mind.* Edinburgh: Tait, Longman, 1820.

BÜCK-RICH, U. *Ernst Heinrich Weber und der Anfang einer Physiologie der Hautsinne.* Zürcher Medizingeschichtliche Abhandlungen No. 70, Juris Druck u. Verlag, Zürich, 1970.

DAWSON, P.M. The life and work of Ernst Heinrich Weber. *Phi Beta Pi Quarterly*, 1928, *25*, 86.

FECHNER, G.T. *Elemente der Psychophysik.* Leipzig: Breitkopf and Hartel, 1860. English Translation as *'Elements of Psychophysics'* by H.E. Adler. New York: Holt, Rinehart and Winston, Inc., 1966.

GELDARD, F.A. *The Human Senses.* Second Edition. New York: John Wiley & Sons Inc., 1972.

GOUDGE, M.E. A qualitative and quantitative study of Weber's illusion. *American Journal of Psychology*, 1918, *29*, 81.

HENSEL, H. Cutaneous thermoreceptors. In A. Iggo (Ed.), *Somatosensory System. Handbook of Sensory Physiology*, Vol II. New York: Springer - Verlag, 1973

HERBART, J.F. Über die verschiedenen Hauptansichten der Naturphilosophie (1823). IN K. Kehrbach (Ed.), *J.F. Herbart's Sämtliche Werke*, Vol V. Langensalza: Beyer u. Söhne, 1890.

KANTOR, J.R. *The Scientific Evolution of Psychology,* Vol II. Chicago: Principia Press Inc., 1969.

LAMING, D. *Mathematical Psychology.* New York: Academic Press, 1973.

LEWIS, T. *Pain.* New York: Macmillan, 1942.

LUDWIG, C. *Rede zum Gedächtniss an Ernst Heinrich Weber.* Leipzig: Veit & Co., 1878.

MURPHY, G. & KOVACH, J.K. *Historical Introduction to Modern Psychology.* New York: Harcourt Brace Jovanovich, Inc., 1972.

REID, T. *Essays on the Intellectual Powers of Man* (1785). Fourth edition. Cambridge: Bartlett, 1853.

TANNER, W.P. Jr. Application of the theory of signal detectability to amplitude discrimination. *Journal of the Acoustical Society of America*, 1961, *33*, 1233.

WEBER, E. H. [Catalogue of scientific papers] *Almanach der kaiserliche Akademie der Wissenschaften,* Vienna, 1852, Jahrg. 2, 203

WEINSTEIN, S. Intensive and extensive aspects of tactile sensitivity as a function of body part, sex, and laterality. In D.R. Kenshalo (Ed.), *The Skin Senses.* Springfield, Illinois: C.C. Thomas, 1968.

THE MAIN WORKS OF E.H. WEBER

Books

Anatomia comparata nervi sympathici. Leipzig, 1817: Reclam.

De aure et auditu hominis et animalium. Leipzig, 1820: Fleischer.

(With W. Weber) Wellenlehre auf Experimente gegründet. Leipzig, 1825: Fleischer.

Allgemeine Anatomie des menschlichen Körpers. Brunswick, 1830: Schulbuchhandlung.
(This work formed Vol I of Hildebrandt's Anatomie)

De pulsu, resorptione, auditu et tactu. Leipzig, 1834: Koehler.
(This monograph contains separate sections on the pulse and breathing, and on the cochlea; the final section, pp. 44-174, is entitled *De subtilitate tactus* and is the work translated here. This work is usually referred to in its short form as *De Tactu*).

Der Tastsinn und das Gemeingefühl.
(This was originally published in R. Wagner, Handwörterbuch der Physiologie, Vol III, pp. 481-588, Brunswick, 1846: Vieweg. It was published separately by Vieweg in 1849 and 1851, and reissued in Leipzig in 1905. This last is the copy used in the present translation).

Zusätze zur Lehre vom Baue und den Verrichtungen der Geschlechtsorgane. Leipzig, 1846: Weidmann.

Annotationcs anatomicae et physiologicae. Leipzig, 1851: Köhler.
(This work contains 3 volumes in which are reprinted most of Weber's earlier monographs and papers. Vol I reprints *De pulsu*..; Vol 2 contains work on the sex organ, blood-corpuscles and lymphatic system, various aspects of histology, and on the liver; Vol III includes papers on the motion of the iris.)

De motu fasciculorum muscularium locali. Leipzig, 1861.

Papers

He published 5 short papers in *Amtlicher Bericht über die 19. Versammlung deutscher Naturforscher und Aerzte zu Braunschweig 1842;* 12 papers in *Berichten über die Verhandlungen der königlichen sächsischen Gesellschaft der Wissenschaften* between 1846 and 1850, including his 1850 paper on haemodynamics; 18 papers in *Meckel's Archiv der Anatomie und Physiologie* between 1819 and 1828; 6 papers in *Müller's Archiv für Anatomie und Physiologie* between 1835 and 1846; and others scattered among various journals, including the paper on musical instruments in *Poggendorffs Annalen,* 1833. An elaboration of the sensory circle theory will be found in *Berichte der königlich-sächsischen Gesellschaft der Wissenschaften zu Leipzig, mathematisch. physische Classe,* 1852, *4*, 87-105. A partial translation is given in R. Herrnstein and E.G. Boring (Eds.). *A Source Book in the History of Psychology,* Harvard University Press, 1965.

Works edited by Weber

ROSENMÜLLER, J.C. *Handbuch der Anatomie des menschlichen Körpers.* Leipzig, 1840: Köhler

HILDEBRANDT, G.F. *Handbuch der Anatomie.* Brunswick, 1830-32: Schulbuchhandlung.

DE SUBTILITATE TACTUS

DE SUBTILITATE TACTUS

ON THE DIFFERENCE IN TACTILE SENSITIVITY* IN DIFFERENT AREAS POSSESSING THIS SENSE [44]

I intend to report some experiments on touch which were planned and carried out together with my dear brother, Eduard Weber. In this subject, as in many other branches of physiology and physics, there is more profit from the combined effort of several men for a short time than from the lengthy and tedious work of one man.

Everything that arouses a sensation in the touch-organ does so by means of some movement of its structure. Not even heat seems to be an exception to this rule; for it is universally accepted that heat stretches, expands and extends bodies. However, we do not perceive heat by touch in the same way as we perceive light by the eye, though the motion of heat and light come from almost the same laws: that is, we do not immediately perceive the intensity and direction of the rays of heat by touch, but only the same effect of heat on our bodies as could also affect a thermometer. [45] Our perception of the skin change caused by heat expansion is very similar. Objects that cause a touch sensation, then, are able to compress or expand the skin - this is certainly the method of stimulation. In fact the skin is compressed either by movement in the objects touching the skin, which the skin resists; or by movement in the area beneath the skin, which the touched objects resist.

---

* *De subtilitate tactus.* Throughout the text Weber uses the term *subtilitas* to cover both 'acuity' (for difference thresholds in weight discrimination and two-point discrimination) and 'sensitivity' (for the absolute threshold, and for differences in the apparent intensity of stimulation in different parts of the body). He did not make a clear distinction between these meanings, though he works towards it in section [111] and following. (See note on p. 80.) In translating this and similar terms we have attempted to use 'acuity' only in the former sense, and 'sensitivity' as a more general term that often includes both meanings. Weber uses *tactus*, 'the touch sense', in a very general manner to include all the cutaneous senses: we have retained the term 'touch' or 'tactile' in translation.

Surprisingly, however, man is in some way helped by the knowledge of his own bodily movement when he examines objects by touch. When we move our arms in a particular direction intentionally and deliberately, we achieve the best perception of the direction of the object under examination. In this way we gain information about the shape of objects that we handle and touch all over. We also perceive weights more clearly when we feel by kinaesthesis* the muscular or nervous** force necessary to lift the objects.

As a general rule, if we wish to discover the causes of complicated phenomena, and their interplay and laws, we must investigate separately the effects of any individual cause and then consider the joint operation of many causes. Similarly, in this case, we must discover what may be perceived by touch if we do not move the touch-organs.

EXPERIMENTS ON THE CORRECT PERCEPTION OF THE DISTANCE BETWEEN TWO OBJECTS SIMULTANEOUSLY IN CONTACT WITH THE TOUCH-ORGAN

When a man with closed or averted eyes makes a judgement about the separation of two objects touching different parts of his skin, his judgements vary in accuracy. This happens even though the objects are identical in shape and material, and are simultaneously brought into contact with the skin by another person. The reason for the variability is that the touch-organ is not equally acute or well-practised all over. These experiments should be carefully arranged, to avoid error. [46] Firstly, they should be performed at a time when our attention is not distracted by too many other sensations, i.e. particularly in the evening or at night. Secondly, the experiments should be repeated and compared for many subjects. Furthermore, the objects, which should be brought to the touch-organ by someone else, should be identical in material and shape, and should also have the same temperature in the various experiments (or at least not much difference). Finally, the two objects should be moved and placed on the touch-organ at the same time and with the same force, but in such a way that the organ is not at all damaged and affected by pain, which would certainly impair the sense of touch.

---

* *coenaesthesis*. This term is used here in the modern sense of 'kinaesthesis', but on p. 91 it is used for the sense of pain.

** *'vim musculorum vel nervorum'* may mean 'the force of the muscles and tendons'. *Nervus* can mean a nerve or a tendon.

Before we performed the experiments (which will be described below in more detail), we used various methods of running the tests and chose the most suitable.

First of all we used the instrument called *Stangenzirkel** in German. This consists of a long metal beam in the shape of a right-angled prism, with two points fastened at right angles (to the beam)**. Because these points can be moved towards and away from each other in a straight line, the instrument seemed suitable for touching the skin with the two points a certain distance apart. To prevent the points from puncturing and hurting the skin, we fixed a cork stopper (*Korkstöpsel*) (used for closing medicine bottles) on each point. Using these stoppers, we touched various parts of the skin of the forehead, face, neck, back of the neck, chest, stomach, back, hip, upper arm, forearm and hand, and finally the thigh, shin and feet. We compared our judement of the distance between the stoppers with the true distance, the judgements being made while looking away. To see whether the effect would be altered if the object touching the skin were spherical and made of a different material from cork, we fixed droplets of gum lacquer to the points instead of the stoppers, and touched the skin with these droplets at various distances apart.

Finally we chose a metal pair of compasses, so constructed that it was possible to fix the legs immovably at any distance apart. [47] We blunted the points so that the

---

* 'Beam compasses' or 'beam trammels'. A beam compass is normally used for drawing large circles, and consists of a beam with sliding sockets carrying steel or pencil points, as in the sketch below.

** *'Constat hoc e ferro prismatico longo et recto, cui duo apices sub recto angulo impositi et adjuncti sunt'*

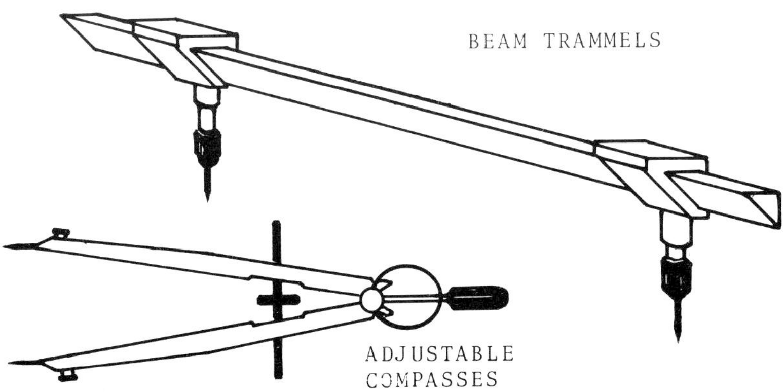

legs ended in a small surface, a third of a *line** in length and width.

With this instrument, then, we touched two spots on the body, a larger or smaller distance apart.

The most appropriate observations to recount now are those that were made and repeated with the latter instrument.

We shall first state the more general propositions that arise from our observations, and then describe the individual findings.

PROPOSITION 1 *The various parts of the touch-organ are not equally sensitive to the spatial separation of two simultaneous points of contact. The sensitivity can be measured by determining the distance between the two points; for it is a property of the touch-organ that it can always distinguish between two points set sufficiently far apart, even if it cannot distinguish between them when too close together.*

If someone else touches our skin with two sufficiently separated compass points while we look away, we are aware of being touched in two places. Moreover, we perceive the position of the points quite accurately, as to whether they are placed longitudinally or transversely across the body; and in several areas we can also make correct judgements about their spatial separation. When the two points are moved closer together, we feel only one impression. The impression is similar to that of a single point, except that it is not perfectly circular - one diameter seems to be longer. In this way it is possible to judge the arrangement of the points: that is, whether they have a transverse or longitudinal orientation. If, however, the two points come closer together, we feel only one impression and do not detect the longer diameter or its orientation.

The finger-tips and the tip of the tongue excel other parts of the touch-organ in the ability to discriminate clearly a small distance separating two objects, [48] but they themselves must be considered almost equal in sensitivity. In these places, if the compass points were two fifths of a Paris *line* apart**, we could distinguish between a transverse and longitudinal arrange-

---

* One third of a Paris line equals 0.75 mm. The pre-metric French system of linear measurement was as follows:

12 points = 1 ligne or Paris line = 0.225 cm.
12 lignes = 1 pouce, thumb, or Paris inch = 2.707 cm
12 pouces = 1 pied or Paris foot = 32.484 cm.
6 pieds = 1 toise = 1.95 m.

** 0.9 mm.

ment[1], even though we did not receive two clearly distinguishable impressions from the two legs of the compasses. When the legs were half a *line* apart* the two impressions of the legs were clearly distinguished, equally well by the finger-tips and by the tip of the tongue. Discrimination was best if one of the points was turned towards the upper surface of the tongue and the other towards the lower surface , and if the compasses touched the finger-tip with the points facing towards the palmar and dorsal surfaces. We should not fail to mention the fact that the individual legs were not easily perceived as separate when arranged transversely.

The palmar surface at the start of the third finger joint is less sensitive than the tip: for in the former area the impression of single legs was certainly not distinguished if the legs were separated by half a Paris *line*, and they were not very well distinguished when separated by a whole *line*.

The back of the tongue must be considered much less sensitive than the tip. If the compass legs were two *lines* apart when touching the middle of the back of the tongue, with one leg pointing to the root and one to the tip, two impressions were not perceived: the legs had to be moved further apart to three *lines* before longitudinal orientation of the compasses could be recognised and a double rather than a single impression perceived[2]. However, on the back of the tongue the opposite rule holds from the one noted a little earlier for the tip of the tongue and the finger-tip. The same is true in many parts of the human body: the face, the scalp, the neck, the whole arm except the finger-tips, and lastly the foot. For in all these parts, as we shall show below, tactile sensitivity is better when the compass legs are placed transversely on the skin than when they are applied longitudinally. [49] On the tip of the tongue and the finger-tips, on the other hand, the legs are distinguished most easily if one leg touches the dorsal surface and the other the opposite side (whether of the finger or of the tongue). In fact these phenomena are not at all contradictory, as will be shown later.

PROPOSITION II *If two objects touch us simultaneously, we perceive their spatial separation and their arrangement more distinctly if they are oriented along the transverse rather than the longitudinal axis of our body.*

The truth of this proposition can easily be tested by anyone in the middle of the upper arm or forearm. Here the compass legs produce an easily distinguishable double impression, if they are placed two *thumbs* apart on the

---

* 1.25 mm.

arm along a transverse axis. However, if they are applied along the longitudinal axis of the arm they give rise only to a single impression, whether they are two or even three *thumbs* apart.

My propositions are tested more extensively by the many observations I have performed on my own body, which are contained in the table that follows. To these I shall later add experiments performed on other subjects.

The following foreword applies to the table: The compass legs were placed on my skin simultaneously, and as far as possible with equal pressure, and were then removed; my eyes were shut, and I did not know the distance apart of the legs; the legs were so close together that I could not judge their separation at the first attempt, and could not even feel that they were separate.

The legs were then placed on various parts of my body, sometimes transversely and sometimes longitudinally across the body and limbs. If the two legs seemed to produce only one impression, very often (unbeknown to me) only a single leg had touched the skin In this way it was discovered whether I could distinguish between the touch of single or double legs. [50] The same experiment was carried out repeatedly by this method, before it was noted whether a single or double impression was felt, and whether I had correctly judged the orientation of the points as horizontal (i.e. transverse) or perpendicular (i.e. longitudinal). In the following table, whenever a reference is made to a perpendicular or horizontal orientation of parts of the body, I meant the orientation to apply to an upright man: nevertheless, I want to make it clear that I was not standing up when the experiments were performed. It should also be noted that I used the Paris foot, which is divided into twelve thumbs, with a thumb consisting of twelve lines. In addition, I measured the distance between the compass legs from the inner edges and not from the centres of the blunted points.

TABLE

of experiments on tactile acuity in individual parts of the body

Place touched simultaneously with two compass legs	Separation of legs in *lines*	Clarity of two points touching the skin			
		Compass horizontal		Compass perpendicular	
		Two points distinguished	Orientation Perceived	Two points distinguished	Orientation Perceived
Tip of tongue	1 ½	clear	very clear	clear	very clear
Same	1	clear	clear	clear	clear
Edge of tongue 4 *lines* from tip	1	not very clear	not clear	fairly clear	poor
Same	1 ½	not very clear	yes	yes	clear
Edge of tongue 10 *lines* from tip	4	clear	clear		
Edge of tongue 1 *thumb* from tip	1	no	no	no	clear
Same	1 ½	no	poor	no	clear
Same	2	poor	clear	poor	
Same	3	poor	clear	poor	
Same	4	yes	clear	yes	
Edge of tongue 1 ½ *thumbs* from tip	1 ½	no	no	no	poor
Same	2	yes	no	yes	poor

Same	3	no	no	no	
Same	4		no		
Same [51]	5	no	poor	no	
Same	6	poor	clear		
Mid upper surface of tongue over 5 *lines* from tip	1	no	no		
Mid upper surface of tongue 1 *thumb* from tip	2	mainly no	yes	mainly no	not clear
Side upper surface of tongue 1 *thumb* from tip	2	no	not clear	no	no
Mid upper surface of tongue 1 *thumb* from tip	3	yes	yes	no	not clear
Side upper surface of tongue 1 *thumb* from tip	3	yes	yes	no	not clear
Upper surface of tongue 5 *lines* from tip	1	yes	yes	yes	yes
Same	$\frac{1}{2}$	no	yes	no	not so clear
Back of tongue 1 *line* from tip	$\frac{1}{2}$	no	yes	no	no
Lower surface of tongue $\frac{1}{2}$ *thumb* from tip	2		clear		no
Upper lip, red part, left and right side	1	no	poor	no	poor
Upper lip, near middle, to left and right	1	no	fairly clear	no	slightly clearer
Upper lip, mid red part	1	yes	clear	yes	clear

Lower lip, mid red part	1	obscurely	clear	yes	clear
Lower lip, red part to right of centre	1	no	obscurely	no	slightly clearer
Lower lip, red part, to left of centre	1	no	obscurely	no	obscurely
Right lower lip, red part	1	poor	clear	no	no
Left lower lip, red part	1	no	no	no	no
Right side of red part of both lips touched simultaneously	1			seemed 2 *lines* apart	clear
Left side of red part of both lips	1			seemed 1 *line* apart	clear
Both red lips, slightly left and right of centre [52]	1			seemed $1\frac{1}{2}$ *lines* apart	clear
Centre of both red lips	1			seemed 2 *lines* apart	clear
Upper lip, side red part	2	yes	yes	yes	yes
Lower lip, red, side part	2	yes	yes	yes	yes
Upper lip, red, side part	2	yes	yes	yes	yes
Mid lower lip, red	2	yes	yes	yes	yes
Upper lip, side red part	3	clear	clear	clear	clear
Lower lip, side red part	3	clear	clear	clear	
Upper lip, mid red part	3	clear	clear	clear	clear

Lower lip, mid red part	3	clear	clear	clear	clear
Lips touched in red and non-red part	3			wide apart	clear
Back of nose	2	fairly clear	fairly clear	no	no
Tip of nose	2	no	not clear	no	no
Same	3	clear	clear	clear	clear
Skin on cheekbone	4	no	no	no	no
Same	6	no	no	no	no
Same	8	poor	poor	no	not clear
Skin on cheekbone and nearby touched together	8	fairly clear	fairly clear	fairly clear	fairly clear
Skin on cheekbone	9	fairly clear	fairly clear	fairly clear	fairly clear
Same	10	yes	yes	yes	yes
Same	12	yes	yes	yes	yes
Mid forehead	6	no	no	no	no
Same	10	yes	yes	yes	yes
Mid forehead, side	12	clear	clear	clear	clear
Upper back of scalp	8	no	scarcely	no	no
Lower back of scalp	8	no	no	no	no
Upper back of scalp	12	yes	yes	poor	poor
Lower back of scalp	12	yes	yes	fairly clear	fairly clear

Upper back of scalp [53]	18	yes	yes	yes	yes
Lower back of scalp	18	clear	clear		
Ear	9	yes	yes	no	no
Back of neck, near spine	12	yes	no	yes	no
Same	14	yes	oblique-ly*	yes	yes
Neck, almost behind 7th vertebra	14	not always	no	yes	yes
Same	15	yes	doubtful	yes	doubtful
Same	16	yes	doubtful	yes	yes
Same	18	clear	oblique-ly	yes	yes
Mid back	12	not always	wrong	no	no
Same	13	clear	yes	yes	yes
Near upper edge of shoulder blade	18	not clear		clearer	
Same	24	partial-ly	yes	partial-ly	yes
Near lower point of shoulder blade	24	no		no	
Same	27	yes	yes	yes	obliquely
Back near lumbar spine	12	no		no	
Same	15	clear	oblique-ly	not clear	no

* The text has *oblique*, which may mean 'sideways' (horizontally) or, 'obscurely'. The latter is perhaps more likely, since the terms also occurs in the 'perpendicular' column.

Same	18	very clear	clear	not clear	no
Same	21	clear	clear	poor	yes
Sacrum (lower back)	12	no	no	poor	yes
Same	15	clear	obliquely	clear	clear
Same	18	very clear	yes	clear	clear
Side of thorax (chest)	12	poor	yes	yes	yes
Same	15	yes	obliquely	yes	yes
Same	18	clear	slightly oblique-ly	clear	yes
Mid front of neck	12	no		no	
Same	15	no	yes	no	no
Same	18	fair		poor	
Same	21	clear	clear	clear	
Side front of neck [54]	21	clear	clear	very clear	
Mid neck	24	yes	very clear	yes	clear
Front neck near windpipe	24	yes	clear	yes	very clear
Front of breast bone	12	yes	obliquely	yes	yes
Same	15	yes	not clear	poor	clearer
Front of stomach, above navel	12	yes	yes	no	no

Same	15	yes	clear	yes	clear
Side of stomach	12	no	no	no	no
Same	15	poor	obliquely	poor	yes
Same	18	poor	no	far apart[3]	yes
Same	21	clear	yes	very clear	yes
Mid front upper arm	10	no	no	no	no
Same near elbow joint	10	obscurely	obliquely	no	no
Mid back upper arm	10	no	no	no	no
Same near elbow joint	10	fair	fair	no	no
Edge of upper chest muscle	12	obscurely	no	no	no
Same	18	clear	clear	no	no
Same	36			no	no
Mid upper arm, front & back[4]	12	no	no	no	no
Same	14	poor	obscurely	no	no
Same	16	yes	yes	no	no
Same	18	clear	clear	no	no
Same	22	clear	clear	no	no
Same	30			not every-where	not every-where
Same	36			not every-where	not every-where

Same [55]	42			yes	yes
Front & back upper arm, near elbow	10	no	no	no	no
Same	12	yes	yes	no	no
Same	16	yes	yes	obscurely	no
Same	18	clear	clear	no	no
Same	22	clear	clear	not always	obscurely
Same	24	clear	clear	yes	yes
Same	26	clear	clear	yes	yes
Under surface of lower arm	7	no	no	no	no
Same near hand	9	fair	fair	no	no
Same	10	yes	yes	no	no
Same	12	yes	yes	not every-where	not every-where
Same	14	yes	yes	not clear	not clear
Under surface of forearm near upper arm	10	no	no	no	no
Same	12	yes	yes	obscurely	obscurely
Same	14	yes	yes	not clear	not clear
Same	18	clear	clear	yes	yes
Mid front lower arm	12	no	no	no	no
Same	14	yes	yes	poor	poor
Same	18	yes	yes	not every-where	not every-where

Same, over flexed muscles[5]	21	yes	yes	very close together	not clear everywhere
Back of head	7	no	no	no	no
Same	8		obscurely		obscurely
Same	9	not every-where	not every-where	not every-where	not every-where
Same	12	poor	uncertain		yes
Same	14	yes	yes	yes	
Palm of hand	3	obscurely	obliquely	better	better
Same	6	clear	yes	clear	yes
Furrow in palm of hand	3	no	yes	clear	yes
Palmar surface of metacarpus of thumb	2	no	no	scarcely	obscurely
Same	3		obscurely		
Same	4	poor	obscurely	poor	obscurely
Same [56]	5	clear	clear	just	clear
Back of knuckles	6		poor		poor
Same	7	yes	yes	yes	yes
Palmar surface of knuckles	1	no	no	no	no
Same	2	no	scarcely	no	scarcely
Same	3	yes	yes	yes	yes
Palmar surface, 1st phalanx of 2nd finger	1	no	scarcely	no	
Same	2	yes	yes		

Furrow of 1st phalanx, 4th finger	2	not clear	yes		
Furrow between 1st & 2nd phalanx	2	not clear	yes		
Furrow between 2nd & 3rd phalanx	2	yes	clear		
Palmar surface, 2nd phalanx of finger	1		obscurely		obscurely
Same	2	clear	clear		yes
Lower surface of 3rd phalanx	1	clear	clear	clear	clear[6]
In vortex of furrowed lines	$\frac{1}{2}$		poor		less clear
Same	$\frac{1}{3}$		no		no
Tip of 2nd finger near nail	$\frac{1}{3}$		poor		clearer
Skin on tense buttock muscle	12	poor		scarcely	
Same	15	yes	yes	yes	obliquely
Front mid thigh	12	poor	poor	no	no
Same	16	fair	long line*	fair	long line*
Same	18	yes	obliquely	yes	yes
Same	21	clear	yes	clear	yes
Mid shin	12	yes	yes	no	no
Same	16	clear	clear	no	no
Mid calf	12	obscurely	no	no	no
Same	16	obscurely	no	obscurely	no
Same	18	yes	obliquely	no	no

* The text reads *linea longitud.* This may mean *linea longitudinis* (a long line), implying that the orientation was clearly perceived; or it may mean *linea longitudinalis* (a longitudinal line), implying that

Same	24	clear	clear	clear	clear
Upper surface of foot near toes	8	poor	poor	no	no
Same	10	not every-where	not every-where	obscurely	obscurely
Same	12	yes	yes	not every-where	not every-where
Same	18	yes	yes	obscurely	obscurely
Sole of metatarsus of big toe [57]	$4\frac{1}{2}$	no	no	no	no
Same	5	poor	obscurely	no	no
Same	6	yes	yes	yes	yes
Upper surface of big toe	$3\frac{1}{2}$	scarcely		no	
Upper surface of smaller toes	$3\frac{1}{2}$	no		no	
Tips of all toes	3	no		no	
Same	4	yes		fair	

I have included this table in order to show the general nature of tactile acuity all over the body. I intend to describe in another table the nature of the touch-sense in particular parts of the body.

The table* contains results for tactile acuity in various parts of my body. The names of the parts are given in almost the same order as the experiments were carried out - an order chosen freely by me. That is why the parts with duller acuity sometimes precede and sometimes follow the more sensitive parts. After I noted for the individual parts of the body the separation of the compass points that allowed correct judgements of their orientation and of the gap between them, I arranged the parts in order of tactile acuity. I then began to repeat the experiments in that order, comparing those parts that were similar or equal in tactile acuity, thus correcting the order. In this way I drew up a table** to show to learned men, in which the separation of the compass legs that was correctly perceived in any part of the body was represented by horizontal lines. It seemed to me that this table showed almost at a glance that tactile acuity differed in various places, and would demonstrate this to others. The shorter the measurements, the finer the tactile acuity; and the longer the measurements, the duller the acuity. However, this is not relevant to the question of thermal sensitivity.

TABLE [58]

of the degree of tactile acuity in the main parts of my body. I measured this as the minimum separation of the compass legs at which I could feel whether the orientation was perpendicular or horizontal, and detect the gap between the legs.

Place	*Separation in Paris lines*	
Tip of tongue	½	-
Palmar surface, 3rd phalanx of fingers	1	–
Red skin of lips	2	—
Palmar surface, 2nd phalanx of fingers	2	—
Dorsal surface, 3rd phalanx	3	——
Tip of nose	3	——
Palmar surface of metacarpal joints	3	——

* i.e. the previous table.

** This table corresponds to the one on p. 199 of *Der Tastsinn*.

Mid-back of tongue, 1 *thumb* from tip	4	——
Non-red part of lips	4	——
Edge of tongue, 1 *thumb* from tip	4	——
Tip of thumb	4	——
Tip of big toe	5	———
Skin covering cheek muscle	5	———
Dorsal surface, 2nd phalanx of fingers	5	———
Palm of hand	5	———
External skin of eyelid	5	———
Mucous membrane, mid-hard palate	6	————
Skin in front of cheekbone	7	—————
Under surface, metatarsus of big toe	7	—————
Dorsal surface, 1st phalanx of fingers	7	—————
Dorsal surface, metacarpal joints	8	——————
Mucous membrane of lips near gum	9	———————
Skin behind cheekbone	10	————————
Lower part of forehead	10	————————
Back of heel	10	————————
Lower part of scalp	12	—————————
Back of hand [59]	14	——————————
Neck under lower jaw	15	———————————
Top of head	15	———————————
Kneecap, and thigh near kneecap	16	————————————
Sacrum (lower back)	18	—————————————
Top of shoulder blade, and upper arm near same	18	—————————————
Buttock muscle, and thigh near same	18	—————————————
Upper and lower part of forearm	18	—————————————
Skin near knee and near foot	18	—————————————
Dorsal surface of foot near toes	18	—————————————
Breastbone	20	——————————————
Spine, towards upper 5 vertebra	24	———————————————
Spine, on upper neck	24	———————————————
Spine, lumbar and lowest pectoral region	24	———————————————
Spine, mid-neck	30	—————————————————

Spine, mid-back	30	________
Mid-upper arm, except where girth of muscles is widest	30	________
Mid-thigh, except where girth of muscles is widest	30	________

PROPOSITION III *In those parts of the body where the two-point threshold for simultaneous presentation is small, the separation of the compass points is perceived to be greater than in parts that do not so excel in the sense of touch.*

Experiments were carried out on the face and hand to prove this proposition. The touch-sense in the middle of the lips is found to be more sensitive and acute than in the side of the lips, near the corner of the mouth. The lips are also more sensitive than the cheeks, which in turn are more sensitive than the skin over the jaw muscles. That is the reason why the legs seem to be together or scarcely separated on the jaw, but farther apart on the cheeks, much farther apart on the side of the lips, and most of all in the middle of the lips. This happens if we touch a man's skin repeatedly over the left jaw muscle, with the compass legs arranged perpendicularly 9-12 Paris *lines* apart, and move the compasses (touching the face lightly) over the left cheek and lips towards the right. [60] As the compasses are moved from the side of the face to the middle of the lips, they seem to grow farther and farther apart; but as they are moved from the middle of the lips to the side of the face, they seem to shrink closer and closer together. These sensations are not peculiar to particular subjects: I discovered them in all the many subjects whom I tested. The statement about the face and lips also applies to the mucous membrane of the tongue. Anyone can verify this by putting the compass legs one or two *lines* apart at the edge of the tongue, and then moving it from the left side to the tip, and from there to the right side, paying attention all the time.

We notice the same thing if we place one leg on the lower jaw and one below it, and then move the compasses to the lower edge of the jaw as far as the chin: the legs then seem farther apart the nearer they are to the middle of the chin. The same thing happens in the hand and fingers, if the compasses (with their legs 4 *lines* apart, for example), are placed transversely on the palm of the hand, and are moved up the three joints to the finger-tip. In this case, tactile sensitivity is greater in the first finger-bone than in the palm of the hand, and increases even more in the second and third finger-bones.

In the same way, the legs of the compasses seem to grow farther apart the nearer they are moved to the finger-tips.

PROPOSITION IV *The compass legs are much more clearly distinguished and seem farther apart if they are placed on two contiguous parts, which are brought together by voluntary movement, than if they touch a solitary part.*

The truth of this proposition can be proved by moving the compasses to the lips, to adjacent fingers and to the two eyelids separate by a narrow chink.
[61] The compass legs, for example, give a single impression and are not discriminated when placed on one lip half a *line* apart; but they are discriminated well if they touch both lips simultaneously. When the legs are so far apart that they are also clearly distinguished by one lip, they seem to be much farther apart if they are placed on the mouth touching both lips at once.

PROPOSITION V *Two compass legs touching us at precisely the same time are distinguished more clearly if they touch two body surfaces that lie opposite each other, or that differ in structure or use, than if they touch only one surface. This rule also holds if one surface has greater tactile acuity than the other. For the compass legs are more clearly perceived if they touch both surfaces than if they touch only the surface with greater acuity.*

I tested this frequently on the lips, both on my own and others'. The compass legs seem farthest apart if one leg is placed on the internal surface and the other on the external surface of the red part of the lips. However, my many repeated experiments confirmed that the internal surface facing the gums is less acute than the external surface.

The same rule holds for the two external surfaces of the lips, the red and non-red parts. The impressions made by the compass legs were clearer and more distinct if they were produced on both parts together rather than on just one of them. When touching both parts simultaneously, the compass legs could be discriminated when they were only 1 *line* apart. The same legs, however, made only one impression if they were placed only on the red part or only on the non-red part. The same thing is found on the edge and tip of the tongue. The legs are most clearly distinguished if one touches the dorsal surface and the other the opposite surface. The compass legs, when half a Paris *line* apart, are distinguished much more clearly on the tip of the tongue if they had a perpendicular rather than a horizontal orientation. [62] Here I must recount some experiments that I also performed on the tips and edges of the fingers. A very small separation of the compass legs is more clearly perceived if they are simultaneously touching the palmar and dorsal surface of the finger-tips than if they touch only the palmar surface. I do not wish to talk about the edge of the hand and foot: anyone can easily prove by experiments what I have reported about these parts of the body. However, I cannot pass over in silence the difference that arises because one part of the skin receives branches from the nerves of the right side and the other from those of the left. This is the reason, in my opinion, why the compass legs are more easily discriminated if they are placed on the back so that one leg touches the right side and the other the left.

PROPOSITION VI *If we investigate in detail the variability of tactile sensitivity in different parts of the body, we perceive that it not only differs in the main parts of the body, but individual spots in the same locality also show differences. They are arranged so that the more acute touch spots are often next to slightly less acute ones. However, these spots do not differ much in tactile acuity, and I have not detected a rule governing the distribution of the most sensitive spots.*

It is not worth stressing any further that tactile acuity is not equally distributed in all parts of the body. This fact seems to indicate that the nervous matter*, on which the sensitivity of the touch spots depends, is also unevenly distributed.

PROPOSITION VII *If one compass leg presses more forcefully than the other, the double impression is more difficult to discriminate. This is because the stronger impression obscures the weaker.*

PROPOSITION VIII *We can discriminate the two impressions more easily if the two legs do not touch us at exactly the same time,* [63] *but with at least some very brief interval interspersed.*

In the Tables that I have presented there is a very general indication of those places where the separation and orientation of the legs is perceived more clearly, and where the horizontal orientation is recognized better than the perpendicular. Those experiments were in fact

---

* *'medullam nervorum'* – *'medulla'* is normally the marrow of a bone, but it can be used for nervous matter (cf: medulla oblongata), or for the innermost or most essential part of any thing.

done along only some lines of the body, for example the mid-line of the spinal column. In order to give a wider conclusion on the sensitivity of the parts, we must now go through them one by one, and state in words the contents of the twofold table*, adding further notes at the same time.

THE TOUCH SENSE OF MY LIMBS

Among the parts of my body with the lowest tactile acuity are the places on the thigh, upper arm, lower arm and leg where the muscles are thickest - the skin covering the middle fleshy part of the biggest muscles. This place is not always exactly in the middle of the limbs, but is never far off. In the front of the upper arm the place is located in the middle of the biceps and inner brachial muscle, and so is nearer the elbow joint. In the calf of the leg it occurs in the middle of the calf muscles (gastrocnemius and soleus), and so is nearer the knee. The upper arm is slightly less sensitive than the lower arm, and the thigh than the leg, but the lower arm is more sensitive than the leg. The fingers and toes are much more sensitive than the middle of the upper and lower arm, thigh and leg. This is particularly true when the compass legs are arranged perpendicularly rather than transversely, and we wish to judge their orientation and separation. As for the rest, the convex part of the joints is more sensitive than the concave part. For example, the kneecap is superior to the back of the knee joint, the outer elbow to the inner fold of the elbow, and the top of the shoulder to the armpit.

To make this clearer, I shall give an example:

When the compass legs are 16-18 Paris *lines* apart on the shoulder, elbow, buttock and kneecap, the perpendicular and horizontal orientations are recognized equally well, and the separation of the legs is also perceived. [64] However, the same discriminations cannot be made over most of the upper arm, a small part of the forearm, and a very large part of the thigh and leg. As the compass legs are moved gradually apart, so the area in which they can be discriminated grows and extends from the fingers and toes to the middle of these parts. Finally, if the compass legs are 2½ or 3 *thumbs* apart, both the horizontal and the perpendicular orientations are recognized over the whole of these parts, and two impressions are felt. The inner skin of the upper and lower arms does not differ much in tactile acuity from the outer skin, and neither does the skin on the back of the thigh and leg from the front skin. Sometimes the posterior skin of the thigh and upper arms seems less sensitive.

---

* *'tabula duplici illa'*. Weber is presumably referring to both of the previous tables, which contain similar information.

If we are considering tactile acuity, the upper and lower arms, thigh and leg must be placed far behind the fingers and toes; and the toes are inferior to the fingers by a surprising extent.

As regards the *hand*, acuity on the palm is much finer than on the back, and becomes even finer the nearer the spot is to the finger-tips. The palmar surface of the fingers (metacarpals) is less acute than the skin covering the knuckles. However, the latter region is inferior to the skin of the first finger-bone, which in turn is inferior to the second finger-bone. The third finger-bone is superior to all these places, and has maximum acuity at the finger-tip.

In the *foot*, the tips of the toes do not have such great superiority over the other parts, and the heel is more sensitive than the middle of the sole.

The upper and lower ends of the thigh and shin, upper arm and forearm, have better acuity than the middle part. Similarly, the upper and lower ends of the bones of the hand (metacarpals), and the front and back ends of the part of the foot composed of tarsals and metatarsals, are more acute than the middle part. [65] It follows that the middle of the dorsal surface of the hand, and the middle and posterior part of the dorsal surface of the foot, are less acute than the lower end of the forearm and leg, although generally the forearm and leg are less acute than the hand and foot. For the rest, the lower part of the leg (i.e. the region of the outer ankle and the region where the Achilles tendon crosses the heel) has better acuity than the upper part of the leg near the kneecap; and the lower part of the forearm near the wrist has better acuity than the upper part near the hollow of the elbow. The same applies to the acuity of the ends of the metacarpals and metatarsals near the fingers and toes, in comparison with the ends near the forearm and leg. However, it is not true for all parts of the limbs that the ends nearer the fingers and toes have better acuity than the ends near the trunk. For example, the skin at the head of the bone of the upper arm, covered by the deltoid muscle, seems to have better acuity than the region of the upper arm nearest the elbow joint. A certain variability in tactile acuity is evident among the individual fingers and toes. This should be investigated with more precise experiments at some other time. Finally, from the first table anyone can see that in all parts of the limbs the horizontal orientation of the compass legs is much more easily and clearly perceived than the perpendicular orientation. This variation due to the orientation of the compasses also occurs in many parts of the head; but in many parts of the trunk, as will be shown below, the orientation has a different effect.

THE TOUCH SENSE IN VARIOUS PARTS OF THE HEAD

Among the parts of the head, the skin covered with hair takes first place: it has better tactile acuity than the neck. However, the whole scalp is not equally sensitive. The skin at the crown is slightly less sensitive than the skin near the forehead and temples and in the lower part of the back of the head. The legs made clearer impressions when placed on lines around the crown or around the lower part of the skull, than on lines leading downwards from the crown. The separation of the legs also seemed much greater in the former case than the latter. [66] The compass legs are perceived more distinctly and discriminated better if they are arranged perpendicularly rather than horizontally. This is equally true of the front and back part of the scalp, and the side part, and also around the crown. Next to the hairy scalp (in so far as it approaches the rather poor tactile sensitivity of the ear) come the forehead, temples and the side of the jaw. Tactile sensitivity in the face is greater the nearer the place is to the tip of the nose or the red part of the lips. The tip of the nose has surprisingly sharp acuity, and the edge of the lips even more: only the tip of the tongue surpasses them. Some people may perhaps think that the eyelids possess a more acute touch sense, since it is agreed that differences in temperature can be discriminated best when the objects touch the eyelids. However, what holds for the temperature sense need not apply to the impressions received from solid objects. The eyelids, then, should certainly not be included among those parts of the face with the greatest tactile sensitivity, but should be considered greatly inferior to the tip of the nose and the lips.

At the edge of the lower jaw and on the cheeks, the compass legs are more clearly discriminated and seem farther apart the nearer they are to the mid-line that bisects the face. On the internal surface of the lips, the touch-sense is duller farther away from the edge; so acuity is poorest near the gums. Pain is easily excited in the gums, but we lack a clear sense of the location of the pricks in this area. Such location-sense as we possess seems to depend more on the transmission of pressure through the dental material to the dental nerve than on the sensibility of the gums. In fact, when the legs touch different teeth, we have some perception of the separation of the legs. The most acute sensitivity, for which the tip of the tongue is so remarkable, exists only in the tip: it is confined to a very narrow area at the end*, not exceeding four to six square *lines*. The sensation is more obscure farther away from the tip, both along the back and

* *arctissmisque terminis circumscriptus est.*

the edge, and deteriorates with increasing distance from the tip. In the middle of the back of the tongue, therefore, and in the place near the hyoid bone, the sensation produced by a touch is fairly dull. [67] The mid-posterior part of the dorsal surface of my tongue is divided by a cleft along its length into two equal parts on the left and right; and tactile sensitivity is greatly reduced towards the front end of the cleft.

The mucous membrane covering the hard palate has a blunter touch sense, but the sense becomes sharper in the soft palate. The external ears (i.e. the cartilage covered with skin) have a dull touch-sense.

THE TOUCH-SENSE IN VARIOUS PARTS OF THE TRUNK

The touch-sense in the trunk is less precise than in the head or limbs. Not only do many parts of the trunk have poor acuity, but no part has such good acuity as the fingers, palm, hand, tongue, lips and nose. Even the breasts, whose nipples are very refined, and which are considered to have good sensitivity, lack an acute sense of touch. Those people who confuse tactile acuity with the ability to perceive a certain titillation or a sharp pain are very much mistaken.

The touch-sense in the trunk is noteworthy in another way: in many places the compass legs are perceived no more clearly in a horizontal than in a perpendicular orientation. The direction of the legs has a different effect in various parts of the trunk. There are, in fact, very few places in which the compasses make a clearer impression when perpendicular than when horizontal. This difference between the trunk and the limbs is worthy of our attention, because the nerves in most parts of the trunk run in a transverse direction, but in the limbs in a more longitudinal direction. The sensation, then, actually seems to change in some respect, depending on the longitudinal or transverse orientation of the nerves. To convince myself of the truth of this I carried out two series of experiments with the help of my brother, the Professor of Physics at Halle*. One series was performed on my own body, and the other on my brother's, by the following metho

Three circular lines surrounding the stomach, chest and neck, and four longitudinal lines in the trunk were selected so that the compass legs could be placed on the skin repeatedly, either on the lines or near them. [68] By this method we discovered the places where the impressions of both legs could be discriminated, and the spots touched seemed more or less separated, and the orientation of the legs could be correctly recognized. In the

* This refers to Wilhelm Eduard, who was Assistant Professor at Halle from 1828 to 1831, before becoming Professor of Physics at Göttingen.

circular lines around the chest and stomach we discovered many facts worth noting; but in the circular line around the neck we did not happen to discover such definite rules about the diversity of the touch-sense. For this reason I shall not record the experiments performed in the latter region.

(1) *The first horizontal line* around the chest was drawn so that it passed through the lowest part of the xiphoid process (breastbone), and through points one and a half *thumbs* below the nipples of each breast. The compasses were placed on the body in the following way:

(a) Sometimes, *so that the the legs always touched two points on this line simultaneously; or, to put it another way, so that the legs were placed transversely on the chest.* In this way, by placing the compasses on many different spots on the line, the tactile acuity of each spot could be discovered.

(b) Sometimes, *so that the legs touched two lines, parallel to the circular line and equidistant from it, and were placed on the chest in a longitudinal direction.*

If the compasses were placed on the circular line so that the legs, when two thumbs apart, always touched two points simultaneously, the following observation was made: there are four places in the circle surrounding the chest where the compass legs are felt more clearly and seem farther apart. Two of the places are located in the middle of the front and back of the trunk, and two on either side. In these places perception is very clear. There are other places between them, where the compass legs are felt less clearly and seem closer together.
[69] If the compasses are placed on the xiphoid process so that the legs straddle the mid front line, both impressions are clearly discriminated and the spots touched seem widely separated. If you touch a neighbouring area, the distance between the legs seems smaller. It seems much smaller still if you move the compasses so that the legs do not simultaneously touch points farther out to the right and left sides of the body, but one of them is placed on the mid line of the anterior skin, or is located behind this line*. A fairly distinct perception in the former place changes to a much less distinct one in neighbouring areas. If you now proceed to touch individual points of the circular line nearer and nearer to the side, you will find a place where the compass legs seem very close together. This place is in between the

---

* *'aut pone hanc lineam collecetur'.* This presumably refers to the mid-longitudinal line of the back of the body.

mid-longitudinal line of the front of the body and the lateral line. If you move the compasses from there to points nearer the lateral line, you will notice that the impressions become gradually more distinct; and the distance between the legs will appear greater the farther to the side you place the compasses. There is, therefore, a place at the side where the compass legs seem farther apart and more distinct in a transverse orientation; but not to the same extent as in the middle of the front or back. If you now move the compasses to individual spots in between the side and mid-back, the space between the legs will seem to shrink again with increasing distance from the lateral line. In this way you will find a place between the lateral line and the mid-longitudinal posterior line (observed by me in the sacrolumbal muscle), where the compass legs again seem barely separated. From there, the nearer you approach the mid-longitudinal dorsal line, the more the space between the legs will seem to grow. The increase will be greatest in the place where the compass legs begin to touch both sides simultaneously, when previously they touched only the right or left sides. [70] So if you come to the mid-part of the back, and place the compass legs so that they straddle the mid-longitudinal line of the back, you will notice that the impressions seem very clear and the separation very great.

It is difficult to discover from experiments the precise area of any of those regions where the legs seemed to have maximum or minimum separation, or whether they occupy exactly the same location in all people.

If the compasses were placed on the skin along the longitudinal axis of the body, so that the legs touched two lines parallel to a circular line and equidistant from it, the following was observed: in the circle around the thorax there are four places where the legs are felt less clearly and seem less separated than in other places. Two of them are located on a longitudinal line along the middle of the front and back, and two on either side. In between these places are others where double impressions are discriminated better. The reason is that the tactile discrimination of the skin is generally poorest when the legs have a longitudinal orientation, but becomes very fine for a transverse orientation in the same area. However, it has not yet certainly been shown whether the areas with clearest transverse discrimination and poorest longitudinal discrimination are exactly the same - i.e. whether they have precisely the same location and boundary.

If the compasses are placed on the xiphoid process so that both legs touch the mid-line of the frontal surface, we perceive such indistinct impressions that they seem single rather than double. Sometimes we can judge the orientation of the legs in a different way: the place touched seems to have an oval or longitudinal shape rather than a circular one, so that the orientation of the legs can be inferred from the orientation of the longer diameter. [71] Under the nipple the impressions of the compasses are more clearly distinguished and seem farther apart. On a lateral line, drawn from the armpit to the ridge of the hip bone and located farther out than the breast, the compass legs *are felt more distinctly than in all other parts of the chest,* and seem to have maximum separation. From here to the mid-longitudinal line of the back the space between the legs seems so narrow that, on the mid-line, there is only a single impression - but with an apparently longitudinal shape.

It was sometimes observed that tactile acuity increases irregularly from the mid-longitudinal front line to the lateral longitudinal line, and decreases irregularly from this line to the mid-longitudinal line of the back; but between these lines there is a place where tactile acuity is greater and finer. In my case, for example, a place of this sort is found under the front of the breast towards the sternum, and in the sacrolumbar region. However, I am not sure of this, and prefer to restrict my report to other well-attested facts.

(2) *The second horizontal line for testing the touch-sense of the abdominal skin was selected so that it encircled the stomach transversely, leading through a spot located above the umbilicus and just over one thumb from it.*

> (a) *If the compass legs were placed on this line, so that they were separated by two thumbs while touching the horizontal line, tactile discrimination was found to be almost the same in the abdomen as in the breast.*

The compass legs seemed more divergent and farther apart, and also felt clearer, in one particular arrangement on the circle. This occurred when the legs were moved to the circular line so that they touched the left and right sides of the body simultaneously and at equal distances from the mid-line of the body, first in the anterior and then in the posterior or dorsal part of the abdomen. If the compasses were moved from the mid-anterior place slightly towards the sides, the separation of the legs seemed to decrease; but when they approached close to the sides of the body they seemed to grow farther apart again. [72] On the longest muscle of the back, they seemed to coalesce in one point; but on the mid-longitudinal line of the back they were perceived most clearly, and seemed wide apart.

> (b) *When the compass legs were placed longitudinally on the abdomen two thumbs apart, so that they touched two circular lines parallel to the one described and equidistant from it:*
> When the legs were moved in this way to the mid-line

of the abdomen, they seemed to have a very small separation; and on the mid-line of the back they seemed to converge into one point. Towards both sides the legs seemed farther apart. From these experiments I think the following can be concluded about the abdomen: when the legs are arranged transversely they are perceived and distinguished clearly, though when arranged longitudinally in the same place they are perceived with minimal clarity. On the other hand, a longitudinal orientation produces a clearer percept and discrimination in the places where a transverse orientation produces a less clear percept.

Finally, I must report the experiments performed by me and my brother on lines drawn along the longitudinal axis of the trunk.

The first longitudinal line along which I tested the degree of tactile acuity in the trunk, on the mid-anterior skin, *was drawn from the chin to the neck and chest, over the umbilicus, as far as the symphysis of the pubic bones.*

If the compass legs were placed transversely on this line, so that they made a right angle with the mid-line of the body, their impressions seemed clearest near the chin, and the space between the legs seemed largest; in the neck tactile acuity diminished more and more towards the shoulder bone; but on the sternum it increased again, so that the legs seemed clearer and farther apart than in the middle and lower part of the neck, though they were not as clearly felt, or apparently as greatly separated, as near the chin.
[73] On the lower part of the chest, the xiphoid process and the upper abdominal region, the distance between the compass legs seemed less than in the upper part of the chest. On and below the umbilicus it increased a little, while on the pubic bones it seemed diminished.

These experiments show that the whole mid-anterior line of the trunk, from the mid-line of the face and chin, has by far the best tactile acuity; and most parts of the line possess a uniformly good acuity; but in the highest part of the neck and chest, and the lowest part of the abdomen below the umbilicus, tactile acuity is a little better than in the parts of the trunk in between.

If the compasses were placed on the mid-anterior line of the trunk, so that both legs touched the line simultaneously, the legs seemed far apart near the chin; but on the neck and the upper part of the sternum they seemed slightly closer together; and on the main body of the sternum slightly farther apart (but closer together than near the chin); on the xiphoid process closer together; on the umbilicus farther apart; and finally, below the umbilicus the distance between the legs vanished entirely, so that the pricks seemed to coalesce into one spot.

The second longitudinal line along which I tested the degree of tactile acuity in the trunk, was the mid-

line of the back, descending from the nape of the neck to the coccyx. If the compass legs were placed transversely on this line, so that the line joining the legs made a right angle with the mid-line, I found that the distance between the legs seemed largest under the occiput, and between the gluteal muscles on the coccyx; but on the back it seemed very small. This apparent separation of the compass legs seemed to increase greatly from the sacrum to the coccyx, but to decrease greatly from the nape of the neck to the thorax.

If the compasses were placed on the mid-line of the back so that both legs touched the line simultaneously, a similar effect was observed as when the compasses were arranged transversely. For it was discovered that, on the nape of the neck under the occiput, the separation is greater towards the middle part of the neck, and becomes gradually smaller in the upper part of the thorax; [74] i.e. it increases between the spines of the scapula, and then decreases in the thorax so that the pricks unite; and it increases again on the sacrum, but especially on the gluteal muscles behind the coccyx and towards the anus.

On the longitudinal lines of the trunk, then, there are many places where the pricks are felt more clearly, and others where they are felt less clearly, regardless of whether the legs are placed transversely or longitudinally. The opposite holds for the circular lines surrounding the trunk. For there, whenever the legs are felt more clearly in a transverse orientation, they usually seem less clear when arranged longitudinally - and the reverse. This difference between the longitudinal and transverse lines of the body certainly arises because the different regions of the trunk, where the longitudinal lines were located, possess varying tactile sensitivity depending on the supply of nerves. It is not because the nerves have longitudinal or transverse orientations in the different places; nor because the legs sometimes touch two different nerves and sometimes two parts of the same nerve.

The third longitudinal line of the trunk to which the compass legs were moved, so that tactile acuity might be measured from the clarity of their impressions, *was situated on the side of the trunk.* The line begins in the ribs towards the armpit, descends to the side of the thorax, and comes to an end on the ridge and outer surface of the hip bone. The compasses were placed so that the two legs were equidistant from this lateral line, and the line joining the points was at right angles to it.

When separated by two *thumbs*, the legs seemed farther apart in the uppermost part of this line, under the armpit, than in the lower part; and closer together on the lower ribs than in the upper part; and farther apart between the lowest rib and the ridge of the hip bone than on the lower ribs. On the ridge of the hip bone and on the iliac bone itself, the legs seemed much closer together than above the ridge, but still gave a very clear impression. [75]

To make these observations more reliable, I used more than one method. In the first method I put the compass legs on single points of the skin, then picked them up and moved them to adjacent points, and compared the impressions made in various places. In the other method, I put the compasses on the skin and moved them steadily without interruption to various different places, or asked for them to be moved over my own skin. If this is carried out carefully, the separation of the legs usually seems smaller in less sensitive areas, and larger in more sensitive areas; and so when the compass legs are moved smoothly, they seem to describe two *curved* lines rather than two *parallel* lines, sometimes converging and sometimes diverging.

ON THE CAUSES OF THE VARIATION IN THE TOUCH-SENSE IN DIFFERENT PARTS OF THE BODY.

First I shall demonstrate that *the cause of the variation is not due to the fact that some parts of the skin are visible while others are hidden.* You might suppose that we perceive and discriminate touched spots more clearly, and judge their separation more accurately, in those parts of the skin that we have often inspected, and where we have used vision to gain accurate knowledge of the location and separation of the individual points. This would probably be because the touch sense seems to improve if we frequently compare tactile and visual impressions of the same thing. However, this opinion is not supported by my findings. The mid-back of the hand, which we inspect so often and know from every angle, is much less sensitive than the back of the fingers and the palm of the hand, and also than the lower forearm, which we cover with clothes. The same is true of the dorsal surface of the foot and the outer ankle bone, or the area of the Achilles tendon: the dorsal surface has less tactile acuity than the latter places. What should I say about the fact that tactile acuity is better in the sacrum, and especially around the coccyx, than in most other parts of the trunk? [76] We cannot inspect this region with our eyes! If vision improves the touch-sense, those parts of the body that we inspect frequently should have finer acuity. However, my research confirms that those parts of the body that we never see are often more sensitive than the parts we see every day. Similarly, the area below the chin is much more sensitive than the skin covering the sternum and abdomen. However, there are certainly many places on the front of our body that are seen either by the eyes alone or with the help of mirrors and that possess better tactile acuity than parts on the opposite, dorsal surface; for I showed that the face is far more acute than the back of the head, the chest and abdomen are slightly more acute than the back, and the calf of the leg is less acute than the front of the leg. However, this seems to be due to the structure of the skin rather than to vision. Our body is built so that it moves forwards more easily than backwards or sideways; and it also has most of the sense organs - eyes, nose and tongue - located on the front surface, with the ears turned just a little back. Perhaps touch, like the other senses, seems more refined in the anterior than the posterior parts of the skin. However, this rule is not without exception. As I mentioned earlier, we have better tactile discrimination in the region of the coccyx than in the region of the pubic bones; but the pubic bones are visible and at the front of the body, while the coccyx is at the back. Nothing further need be added to show that vision does little to improve tactile acuity. Besides, this view is confirmed by the very refined sense of touch in the caecum. Acuity is excellent there without the help of the eyes, because our attention is entirely directed towards tactile impressions.

Moreover, *the main reason for large variations in tactile acuity in different places should not be sought in some mechanical aid, such as some parts of the skin being easier to touch than others*; for example, [77] in the fact that some parts of the skin are supported by bones, so that the impression is stronger and clearer. The tip of the tongue is very flexible, but has better tactile acuity than other parts of the skin; and the lips are among the most outstanding tactile organs, but they are not supported by bones.

The cause of tactile acuity, then, is to be sought in the peculiar structure of the organs serving the touch-sense; but we lack more accurate knowledge of this.

The structure is not related to the number of papillae. For there are certain places, e.g. the tips of the breasts, which contain many large cutaneous papillae, but whose acuity is only average. The tongue possesses the most acute touch-sense only in the extreme part of the tip, though the whole upper surface is covered with papillae; yet the papillae of the tip are not obviously different from those at the back. For this reason I maintain that the variation in the touch-sense depends upon the density, pathways and endings of the nerves, which are different in different places.

ON THE GREAT IMPROVEMENT OF TACTILE ACUITY PRODUCED BY MOVEMENT OF THE TOUCH-ORGANS AND OF THE TEST OBJECTS

At the beginning of this report I mentioned that movement of the touch-organ is a great help to someone making a tactile inspection of objects. This is undoubtedly true. Indeed, the shape and texture of the objects is not discovered by touch, unless the finger is deliberately moved over the surface of the test object.

If, for example, the points of the compasses were so close together that you perceived only one prick, you would nevertheless feel the separation very clearly if you moved the finger from one point to the other. You will notice the same thing on the palm and back of the hand, and in other places. The main cause of this effect is as follows: we notice the time that elapses between the touch of the first and second points on the same part of the finger; and we are aware of the speed with which we move the finger; and we then measure the space between the points from the time passing between the two pricks. [78] In this way, by moving the finger, we discover very finely spaced irregularities, which we certainly do not feel if the finger is kept stationary on the irregularities.

The separation of the compass legs also feels slightly clearer if the legs are moved over our skin by someone else. In this experiment, the legs touch and depress various spots of the skin. The perception becomes clearer with this method, because the touch of the compass points is repeated very frequently and rapidly in different places. We estimate the distance between the legs from several tactile stimuli rather than one, and merge several perceptions into one.

For the same reason, immobile objects are perceived more clearly if the hand or fingers are drawn over them. In this case also, the same object makes an impression in many parts of the hand and fingers. The second reason, then, for the clearer tactile recognition of objects with a moving hand, is that the same impression is repeated in several parts of the skin.

There is a third reason why tactile recognition improves with movement of the fingers or hand. When we move our hands or fingers we are certainly aware of their orientation in relation to ourselves, and this makes it possible to judge the shape of the object over which they are moving. The object forces us to change the direction of the hand, so that the object is not in the way of the movement.

No one should be surprised that we are aware of the direction of movement of our arms and feet in relation to ourselves; for we see these movements every day, and if we do not see them we still know them by touch. Many people may perhaps be surprised that we also perceive the direction of such movements very clearly when we have never seen them and when they are caused by muscles that are very small and little used. I shall illustrate this with an example. The bulb-endings of the hairs register sharp pain if the hairs are pulled, and sense the movement of the hairs when disturbed by the wind or other very light cause; but, in addition to this, they are compared so that we make completely accurate judgements about the direction in which the hairs are pulled. [79] If, for example, someone grasps a bunch of my hair, and pulls and stretches it slightly in a certain direction, I can indicate the direction of the pull very accurately, with my eyes shut, by pointing my finger.

If someone believed that this knowledge was derived only from the touch of the bulbs, he would certainly be wrong. I shall show instead that the source of the knowledge lies in the muscles that move the head and the scalp. The direction of the pull is recognized most clearly if the movement of the head is unrestrained. In this case, the force stretching the hairs easily initiates movement of the upright head, but the muscles holding the head erect resist this motion, and cancel it with an opposite motion. Our mind, which is aware of this muscular movement, makes a correct judgement about the direction of the motion communicated by the hairs. However, if someone else grasps my head and holds it immobile in a certain position, I cannot perceive the direction of the hair-pull as clearly as with my head free and upright.

Nevertheless, the direction is still perceptible; but the feeling is derived from the movement of the muscles, as they move the scalp. If you also prevent the scalp moving when the hairs are pulled, then you completely remove any sense of directional pull. This is the way to prevent the scalp from moving: press the skin to the skull, and immobilise it in those places where skin movement usually arises when the hairs are pulled. If the head and scalp are not moved at all when the hairs are pulled, there is no reason why the muscles should resist the movement of the hairs; so, in this case, we cannot tell the direction of movement. I also performed experiments on other people, similar to those I reported a little earlier, with exactly the same result. It is fairly obvious that the tactile perception of objects is greatly enhanced if the touch-organ is moved in a deliberate and appropriate manner. It is not surprising, then, that the tactile recognition of objects is very poor if the test objects are moved over stationary touch-organs. [80] Shut your eyes, hold your hand still, and ask someone to bring various objects to your finger-tips - objects such as a piece of paper, glass, sheets of metal, smooth wood, leather, linen, smooth silky fabric and rough fabric, and other unknown objects. You will certainly be surprised at the imprecise quality of the sensation, which does not enable you to recognize the nature of the objects. If the objects are pressed to the finger with varying force, you will often consider identical objects as different, and very different objects as identical.

When some metal object is pulled over our finger-tips by someone else, we deduce that it is made of metal from the sensation of coldness and hardness; we have this sensation because metal objects, being at air temperature, quickly absorb heat when they touch part of the skin. If metal has exactly the same temperature as our skin, it is scarcely distinguishable from glass or other objects; for the other common characteristic of a metal - a very high specific weight* - is not recognized if someone else touches our hand with the metals, so that we do not feel their weight. If our stationary finger is passively immersed in a liquid substance such as water or mercury, with someone else bringing the flask of liquid to our finger, we recognize its liquid nature from the fact that it presses the finger only slightly, and makes contact all the way round the finger simultaneously. If someone presses a flat sheet of glass or metal to our stationary finger, pressing only lightly at first, and then more strongly, and then lightly again, the object seems to have a convex surface; on the other hand, if the order of stimulation is reversed, it seems to have a concave surface.

It is certain that the organ of touch becomes more sensitive with practice: medical obstetricians, for example, when examining the reproductive organs of pregnant women and women who have just given birth, can feel and distinguish much more than less practised men. [81] However, it seems uncertain whether the sensitivity of the touch-organ increases because it is nurtured and perfected by frequent exercise; or whether, instead, people learn to make better use of this organ: they become used to disregarding other senses and thoughts and concentrating only on touch, and they learn to move their limbs more appropriately and to reach firmer conclusions about the nature of the objects from the impressions they receive. The latter opinion seems much more likely to me. If touch were improved by nurture, then tactile acuity would be outstanding in practised people even if the objects were moved to their immobile hands; but if tactile acuity resided in the appropriate use of the organ rather than in its composition, I think it would be observed only with movement of the limbs.

ON TACTILE SENSITIVITY IN DISCOVERING THE WEIGHT OF OBJECTS

There are four most important aspects of objects which we discover by touch:

(1) the force with which they resist the pressure of our organs, and

(2) their shape and the shape of the space between them

---

* *pondus specificum*, or 'specific weight' is the weight per unit volume. 'Specific gravity' - the more familiar current term - indicates how many times a certain volume of the substance is heavier than an equal volume of water.

(3) the force with which they press our organs, and especially their weight

(4) lastly, their temperature, whether hot or cold.

It is agreed that the tactile tense is excellent, whatever method it uses for discovering the complete shape of an object, and the shape of the surface of its individual parts, and the shape of the space separating the objects (i.e. the distance between them): it is considered good regardless of the variation in accuracy of these judgements, which is due to the diversity of the touch-organ.

I must now report the results of my experiments on the sensitivity of the touch-sense in discovering the force with which objects press the organs, and particularly on weight perception.

The weight of an object is perceived in two ways: first by the touch-sense in the skin, and then by the special* sense of the voluntary muscles. The latter sense tells us the degree of tension of the muscle when lifting weights and other objects.

[82] These two methods of discovering the weights of objects are very different: the former method depends upon the objective sense of touch, while the latter depends on the subjective sense of muscular kinaesthesis. This assumes, of course, that we call a sense 'objective' when we use it to perceive objects that have a certain pressure on our organs and produce some effect; and that we call it 'subjective' when we seem to perceive only the effect of the objects and not the objects themselves.

To determine whether weight perception depends more on touch or on the muscular sense, and which parts of the touch sense are used, experiments must be arranged so that only the touch organs, and not the muscles are used.

For this purpose the hands, or other tactile organs, are placed on a seat or wooden table so that they remain still when a weight is placed on them; but they are secured in such a way that muscular effort is not required to hold them in the right place. If you place unequal weights on a subject's immobile hands, and he does not know which hand supports the heavier weight; and if you vary one of the weights without his seeing; then, with a sufficient difference in weight, he will distinguish the heavier weight consistently, even when the hands are changed over.

You will appreciate the acuity of the touch-sense by the difference between the weights. If the subject discriminates only large weight differences, it is obvious that the touch-sense is imprecise; but if he distinguishes very similar weights, you will conclude that the touch-sense is very acute.

---

* *proprius*.

Although weight discrimination is very fine with touch alone, it is always finer if the weight is lifted by the hand. For then the touch-sense is joined by the sense of muscular effort when moving the hand together with the weight.

If you know the size of the weight difference that is sufficient for tactile weight discrimination with stationary hands, you can discover how much the lighter weight can be increased, while still preserving the discrimination, when the hands are used to lift the weights. [83] The additional weight will indicate the extent of the gain in precision when the weights are lifted by the hands.

This method, then, will enable you to measure the precision of the muscular sense in different people - the sense by which we perceive the effort used to move obstacles. The method separates the contribution of the touch organs and the muscles, and shows the rôle of each sense in weight perception.

This method of investigating and measuring tactile acuity and muscular kinaesthesis separately is all the more noteworthy because I cannot see another way in which we can observe and measure the muscular sense accurately.

If these experiments are carried out on the hands, and also on the lips, forehead, back of the head, shoulders, arms, stomach and feet; and if the individual phalanges of the fingers are examined in this way as well as the whole of the hands, and the dorsal surface of the hand and foot are compared with the palm and sole; then conclusions may be reached about the variation in different organs of the sensitivity of the touch-sense for weight estimation.

My experiments reveal great variation in this respect between different parts of our body. The lips are best at correct and precise weight perception; the fingers and toes follow next, with the third phalanx surpassing the second, and the second the first; then follow the palm of the hand and the sole of the foot. However, the back, stomach, shoulders, arms, feet, and back of the head, are much less precise at weight estimation. There seems to be good agreement between experiments, if we compare these experiments with the previous ones on tactile acuity, when we measured acuity through the ability to discriminate the separation of the compass legs on various parts of the body.

ON THE DIFFERENCE BETWEEN THE RIGHT AND LEFT SIDES IN THE ACCURACY OF WEIGHT PERCEPTION BY TOUCH* [84]

It should be mentioned that there is often a difference in the touch sense between the right and left sides. Just as the two sides differ in muscular strength, so, according to my experiments, they also differ in cutaneous sensitivity. For most people, the right side is superior to the left in muscular strength; however, we find the opposite if we examine tactile sensitivity on both sides, by placing the weights on various parts which are kept still and immobile. The same weight seems heavier to most people when placed on the left side, and lighter on the right side; and different weights often seem equal if the heavier is placed on the right and the lighter on the left, while they seem unequal if the heavier is on the left and the lighter on the right. This phenomenon seems to be due to the fact that the sensitivity of the left side is generally finer than that of the right.

This observation was first made on my own body by my friend Seyffarth, Professor at Leipzig**. Seyffarth once gave me advice when I was researching into the nature and laws of waves; and similarly again in this investigation he drew my attention to the difference between the left and right sides. This difference is so large in my body that it shows up not only in the hands but also in the soles of the feet and in other places; and it seriously hinders the precise comparison of weights, when they are placed one on each side of the body.

I am certainly not claiming that the left side always possess a finer weight-sense. For when I placed weights on the hands of many subjects, and examined their touch-sense, I discovered the following: in most people the left hand possesses the finer touch-sense, but in some the right, and in a few there is no difference between the two hands.
[85] These were the results for fourteen people, varying in sex and age, and engaged in various studies and occupations: eleven felt the same weight as heavier when it lay on the left hand rather than the right; for two the opposite was true; while only one felt no clear difference between the two sides.

---

* An alternative translation of the following section [pp. 84-5] by J.D. Mollon appears in Kinsbourne, M. (ed.), 'Asymmetrical function of the brain', New York: Cambridge Univ. Press, 1977 (in press).

** Probably Gustav Seyffarth (1796-1885), the Egyptologist.

PRECAUTIONS TO BE TAKEN IN THIS CLASS OF OBSERVATION

To avoid errors, weights of the same metal and shape should be used. Two equal weights do not produce the same sensation in the skin if they are shaped so that one of them touches a larger area, and is therefore supported by more points of the skin than the other. The cause is obvious. Clearly, the individual points of the skin are pressed with a greater force by the weight touching the smaller area, but fewer points are pressed - or, to put it a different way, the surface area is smaller. Conversely a given point is pressed with less force by a weight that presses over a larger area. The former weight, therefore, produces the same sensation as metals that have an exceptionally high specific weight, such as gold and platinum, while the latter arouses the same sensation as metals with a lower specific weight. However, it is very difficult to compare the absolute weight of two objects correctly, if their specific weight is very different. In our example, the mind cannot easily calculate the force pressing the skin, since the force must be recognized from the area of skin-contact and from the degree of pressure on any point of that skin.

Care should also be taken that metal weights have the same temperature; or, if this is troublesome to arrange, that something should be placed between the weight and the skin - some material that does not easily conduct heat, such as cardboard. The estimation and discrimination of two weights is much poorer if the weights differ in temperature. The mind is distracted by the twofold difference of weight and heat, and is in danger of confusing these different impressions from the two sources. [86] Because of this double stimulus, it is useful to insert a layer of cardboard between the weights and the skin. These boards prevent the different temperatures of the weights from having different effects; and they immediately correct any variation in shape, because you can easily use them to give an identical area.

Next, the weights should be placed on the two hands in such a way that they touch the same part of the hand. The reason for this instruction is obvious from the above account concerning the difference in tactile sensitivity in different parts of the skin. Because the touch-sense is more sensitive in the fingers than in the metacarpals, a weight placed on the volar surface of the fingers seems a little heavier than if it rests on a metacarpal. Even in the individual joints of the fingers this type of difference is found; in fact, the nearer the weight is to the tip of the fingers, the greater its effect.

ON THE MEASUREMENT OF THE RELATIVE CONTRIBUTIONS OF THE TOUCH-SENSE OF THE SKIN AND OF THE MUSCLE-SENSE TO WEIGHT ESTIMATION

I placed two-pound weights on many subjects' hands, which were motionless on a table, and inserted a cardboard sheet between them. Then I reduced the magnitude of one of the weights without the subject's knowledge, and changed the hands supporting the weights - by switching the lighter weight alternately to the right and left, of course. In addition, I often lifted the weights and replaced them on the same hands, so that the subject could not guess, but could feel the side with the heavier weight only by touch. I recorded it only if the subject discriminated the heavier from the lighter weight correctly on repeated attempts and with frequent changes of hand.

The same experiments were next repeated on the same people, but in such a way that they lifted both their hands and the weights together, and weighed the weights by hand. Then, if I discovered the size of the just noticeable difference, I again made a note of it, and compared the numbers representing the difference between the weights.

[87] In this way I found that the weight-differences that were clearly detected were as follows:

			Ratio of heavier to lighter weight
(1)	by a merchant unskilled at weighing objects	1st trial	16:13
		2nd trial	32:31
(2)	by an educated man, a mathematician	1st trial	16:13
		2nd trial	64:59
(3)	by myself	1st trial	2:1
		2nd trial	16:15
(4)	by a merchant unskilled at weighing objects	1st trial	4:3
		2nd trial	8:7
(5)	by a girl	1st trial	16:8
		2nd trial	16:15
(6)	by a woman	1st trial	2:1
		2nd trial	8:7
(7)	by a woman	1st trial	8:5
		2nd trial	15:16
(8)	by a student of literature	1st trial	4:3
		2nd trial	32:29
(9)	by a student of literature	1st trial	8:5
		2nd trial	16:15
(10)	by a student of literature	1st trial	16:13
		2nd trial	64:61

I shall change these numbers into the number of ounces on each hand, so that their meaning is more obvious. This presentation shows the just noticeable difference in ounces, both with stationary hands and with the weights lifted by hand. In this table, for the sake of brevity, I shall always denote the trial made with stationary hands by the letter *a*, and the trial made with lifted hands by the letter *b*.

Number of experiment[88] *	Number of ounces on each hand at threshold	Difference threshold in ounces
(1)	a. 32:26	6
	b. 32:31	1
(2)	a. 32:26	6
	b. 32:29 ½	2 ½
(3)	a. 32:16	16
	b. 32:30	2
(4)	a. 32:24	8
	b. 32:28	4
(5)	a. 32:16	16
	b. 32:30	2
(6)	a. 32:16	16
	b. 32:28	4
(7)	a. 32:20	12
	b. 32:30	2
(8)	a. 32:24	8
	b. 32:29	3
(9)	a. 32:20	12
	b. 32:30	2
(10)	a. 32:26	8
	b. 32:30 ½	1 ½
(11)	a. 32:17	15
	b. 32:30 ½	1 ½
(12)	a. 32:22	10
	b. 32:30 ½	1 ½
(13)	a. 34:16	18
	b. 32:24	8
(14)	a. 32:20	12
	b. 32:26	6

* Nos. 1-10 clearly refer to the subjects of the preceding table, but it is not clear what data Nos. 11-17 refer to.

(15)	[89]	a.	32:26	6
		b.	32:28	4
(16)		a.	32:24	8
		b.	32:31	1
(17)		a.	32:26	8
		b.	32:28	4

Let us now see what the experiments listed in this table show about the precision of the touch-sense at weight estimation.

In the first place, the following fact is made quite clear, beyond all shadow of doubt: *when someone compares the weights of objects by touch alone, with stationary limbs, he does not discriminate the difference as accurately as when he lifts the limbs and weights together.* Most subjects did not perceive a difference less than about eight, ten or twelve ounces, if they compared the lighter weight on one hand with a weight of 32 ounces on the other. Indeed, if the difference was less, the two weights seemed equal. But if the same subjects lifted the weights with their limbs, they perceived a difference of one and a half, or two, or three, or certainly four ounces.

It turned out, then, that subjects observing and estimating weights with the combined help of the touch-sense and muscular kinaesthesis, were able to distinguish a difference equal to one fifteenth of the heavier weight. However, the same subjects usually distinguished a difference of only about one eighth with touch alone. Nevertheless, there are a few other people who possess an outstandingly poor or acute touch-sense. Some have no clear tactile recognition of a difference equal to half the standard weight; while others possess such an acute touch-sense that they can discriminate a difference equal to one sixth of the standard weight, with stationary hands. [90]

It is shown next from the former observations that purely tactile weight-estimation becomes twice as precise with the addition of muscular kinaesthesis.

If we note in round numbers the weight difference that we can only just discriminate when investigating and estimating weights sometimes by touch alone and sometimes by touch and kinaesthesis together, we get the following fractions:*

---

* This table contains very approximate, and sometimes grossly inaccurate fractions derived from the first 13 entries of the previous table.

Number in previous table	Differential threshold for touch alone, when heavier standard was 32 ounces	Same for touch and muscular kinaesthesis
1.	1/5	1/32
2.	1/6	1/16
3.	1/2	1/16
4.	1/4	1/8
5.	1/2	1/16
6.	1/2	1/8
7.	1/4	1/11
8.	1/4	1/16
9.	1/4	1/16
10.	1/2	1/16
11.	1/3	1/16
12.	1/2	1/4
13.	1/3	1/5

[91] However, this conclusion - that our weight-estimation is more than twice as good if we use muscular kinaesthesis in addition to touch - is not worth much if it applies only for a weight of 32 oz., and not for lighter or heavier weights.

I should, then, also mention other experiments, which prove that touch and kinaesthesis have the same ratio to each other when judging much lighter weights as they do when two pounds or 32 ounces are placed on either hand. Using the same subjects as I had tested previously with two weights of 32 ounces, I now placed weights of 32 drachms* on their hands, i.e. one eighth of the former weight. Although I suspected that they would not perceive the difference as clearly for two objects one eighth lighter, the experiments nevertheless showed that the same difference was distinguished equally well by touch for the lighter ahd heavier weights.

I shall present four experiments demonstrating the point. Four subjects (whom I prefer to denote with numbers) compared fairly heavy weights of 32 ounces on their stationary hands, the two weights being equal. I then began to reduce one of the weights until the subjects noticed the difference between them. I noted this difference, and repeated the same experiment with the weights lifted by hand, so that they were estimated with the simultaneous help of both touch and muscular kinaesthesis. I then noted again the difference that they could scarcly detect.

I now presented lighter weights instead, weighing 32 drachms, in exactly the same way; and I made a note of

* Weber is using the Apothecaries' Weight scale, which contains 8 drachms to an ounce. 1 drachm is 3.89 grams.

the differences that were not noticed, evidently escaping attention*.

Now, if you compare the differences for the heavier and lighter weights, which have been subtracted for our consideration, you will notice that they are almost the same.

Subjects' number in the experiments [92]	Smallest perceptible difference when ounces or drachms were placed on the hands			
1. Touch	32 oz.:17 oz.	difference	15	oz.
Touch & kinaesthesis	32 oz.:30½ oz.	"	1½	oz.
Touch	32 dr.:24 dr.	"	8	dr.
Touch & kinaesthesis	32 dr.:30 dr.	"	2	dr.
2. Touch	32 oz.:22 oz.	"	10	oz.
Touch & kinaesthesis	32 oz.:30½ oz.	"	1½	oz.
Touch	32 dr.:22 dr.	"	10	dr.
Touch & kinaesthesis	32 dr.:30 dr.	"	2	dr.
3. Touch	32 oz.:20 oz.	"	12	oz.
Touch & kinaesthesis	32 oz.:26 oz.	"	6	oz.
Touch & kinaesthesis	32 dr.:26 dr.	"	6	dr.
4. Touch	32 oz.:26 oz.	"	6	oz.
Touch & kinaesthesis	32 oz.:30 oz.	"	2	oz.
Touch & kinaesthesis	32 dr.:29 dr.	"	3	dr.

ON THE CAUSE OF THE DIFFERENCE BETWEEN THE RIGHT AND LEFT SIDES OF THE HUMAN BODY IN THE PERCEPTION OF WEIGHT**

The same weight seems, for most people, to press less hard on the right hand than the left. The reason for this is often thought to be the hardness and thickness of the skin, which has been increased by much work and frequent pressure. We may suspect that the sense of touch is blunted in the right hand because most people use that hand more often than the left.

Alternatively, the cause may be sought in the muscular differences between left and right. This explanation depends on the assumption that a weight lifted by hand seems heavier on the side where the muscles are weaker, since they contract with a greater effort to lift the weight.

[93] However, both explanations are wrong. The first is wrong because a thick cuticle does not impair weight

---

* Weber describes the threshold difference interchangeably as 'just noticed' or as 'just not noticed'.

** An alternative translation of the following section [pp. 92-4] by J.D. Mollon appears in Kinsbourne (op. cit. p. 57).

perception. The skin on the heel and the sole of the foot is thicker than anywhere else, and all thermal sensation and two-point discrimination are remarkably impaired; nevertheless, weight comparison is performed there with great ease and precision. The other explanation, based on the muscular weakness of the left side, is not able to explain what we want. For there would be a difference in weight perception only if weights were lifted and not if they were placed on passive hands.

Therefore, since we have no other explanation, it is likely that the difference lies in the structure of the sensory nerves. It is quite possible that the sensory nerves on the left are more sensitive than those on the right, just as the muscles of the right side are thicker and stronger than those of the left.

In this way one can easily explain why weight perception by touch is more sensitive on the left than on the right, for the foot and shoulder in addition to the hand. The foot and shoulder are not experienced in the perception and judgement of weights, and yet they show the same difference as the hands.

Some time ago, on another occasion, I explored the question whether the difference between the two sides that had been noted for muscular strength was also found in the nervous system and in the perceptual faculty*. At that time I thought the eye was an instrument uniquely suitable for carrying out experiments on this subject, and I suggested such a research project to Holke[7]. I had at some time inspected the notebooks of Tauber, an optician who was a Master of Liberal Arts and had formerly been secretary to Hindenberg (Professor of Physics at Leipzig). [94] I saw that Tauber had, over a number of years, examined the eyes of many people who were buying spectacles from him; he had measured their visual acuity and optimum distance for reading, and had kept quantitative records of the difference between the eyes. I hoped that it would be worth the work if a careful comparison of these observations were to show whether both eyes usually enjoy the same acuity at a given distance or whether one eye has better acuity than the other.

It seems, however from Holke's analysis of Tauber's record that there is no obvious difference between the two eyes in this respect.

I was very surprised to find a difference in favour of the left side for the touch organ, since I had failed to find any difference for the visual organ.

I hoped that I would find a similar difference between the two hands by another method - by placing two different weights simultaneously on the palmar surface of

---

* *in systemate nervoso.... et circa sentiendi facultatem.*

the fingers, and noting the smallest difference that could be observed with confidence. I suspected that these differences would be greater for the right hand and smaller for the left.

The following method was used for these experiments. I placed a weight of 32 drachms on the subject's right index finger, on the capitulum of the metacarpal bone, and another slightly lighter weight on the capitulum of the little finger. I asked the subject to compare the weights as accurately as possible, first holding his hand still, then lifting it together with the weight. The same experiment was then repeated on the left hand of the same subject. However, I was disappointed: this method did not reveal any superiority of tactile acuity for the left hand.

ON THE VARYING SENSITIVITY BETWEEN ORGANS, WHICH MAKES THEM DIFFER IN RECEPTIVITY TO WEIGHT IMPRESSIONS

We have two ways of discovering the variation between different organs. The first way is as follows.[95] We place two different weights on the two palms of the hands, the two backs of the hands, the two forearms, shoulders, upper arms, sides of the forehead, and sides of the back. These tests should be performed on the same subject, keeping the organs stationary. We see whether this subject discriminates the same difference in all these places, or whether a greater difference is needed for tactile weight discrimination on the back of the hand than on the palm, and on the forearm than the back of the hand, and so on. Experiments show that weight discrimination varies considerably in different organs. The larger the just noticeable difference, the blunter is the touch-sense; and the converse is also true. With this method, we place the weights *successively* on the two organs whose sensitivity we are testing.

In the second method we place two weights *simultaneously* on the organs. Equal weights do not seem to exert the same pressure on organs that differ in sensitivity: the pressure feels greater where the sensitivity is greater. If you place one of the weights on a less sensitive organ, and reduce its weight until the two feel equally heavy, the difference in weight will then tell you the difference in sensitivity of the two organs.

This method of comparison can, of course, be used only for those organs whose surfaces are oriented so that weights can be placed on them simultaneously.

For example, it is obviously impossible to place two weights simultaneously on the back and palm of the same hand, or on the forehead and the back of the hand. We shall now follow the first method for investigating the different grades of sensitivity in different parts of the skin. In these tests I placed a constant and a variable

weight on some organ of the right or left side.
[96]The constant weight was composed of six silver coins, all of the same size and weight. I arranged six Joachims* (a type of coin) in one column. Ten Joachims contain one pound of silver and almost a sixth of a pound of copper. The variable weight consisted of six, five, four, three or two Joachims. The variable weight was reduced until the difference was discriminated by touch. The constant weight was almost equal to twelve ounces, all Joachims weighing just under two ounces.

If you compare the number of Joachims in the variable weight for different places, you will make a general assessment of the variation in tactile sensitivity; for the smaller the number, the blunter the touch-sense.

In this way I showed that weights made the heaviest impression on the volar surface of the fingers; on the sole of the foot at the capitula of the metatarsal bones and the beginning of the toes; and on the forehead. The same weights made a less clear impression on the concave part of the sole; the back of the head; the frontal surface of the chest; and other places. The heel holds a place midway between these organs.

---

* A Joachim, or *Joachimstaler*, was a silver coin first minted in St. Joachimsthal for King Ludwig I of Bohemia in 1515. The omission of the place-name Joachim gave rise to the coin-name 'Taler' or 'Thaler'. A picture of a Joachimstaler can be found in *Brockhaus Enzyklopädie* Vol. 9 p. 455. F.A. Brockhaus, Wiesbaden 1970.

Places where two weights were placed simultaneously	Number of Joachims placed	Weight if 1 Joachim = 2 ounces	Ratio if 6 oz. has 32 parts	No. of just noticeable parts
Volar surface of fingers of either hand	6 & 5	12 & 10 oz.	$32:26\frac{2}{3}$	$6\frac{1}{3}$
Volar surface of metacarpal of hand	6 & 4	12 & 8	$32:21\frac{1}{3}$	$10\frac{2}{3}$
Dorsal surface of fingers	6 & 4	12 & 8	$32:21\frac{1}{3}$	$10\frac{2}{3}$
Internal surface of upper arms	6 & 2	12 & 4	32: 8*	24*
Sole of foot on metatarsal capitula	6 & 5	12 & 10	$32:26\frac{2}{3}$	$6\frac{1}{3}$
Concave part of sole [97]	6 & 2	12 & 4	32: 8*	24*
Heel	6 & 3	12 & 6	32:16	16
Same when chemically desensitized***	6 & 2	12 & 4	32: 8*	24*
Right and left part of forehead	6 & 5	12 & 10	$32:26\frac{2}{3}$	$6\frac{1}{3}$
Hair-covered back of head	6 & 2	12 & 4	32: 8*	24*
Frontal surface of chest	6 & 2	12 & 4	32: 8*	24*
Either shoulder	6 & 4	12 & 8	$32:21\frac{1}{3}$	$10\frac{2}{3}$
Midline of back, near shoulderblade	6 & 1	12 & 2	32: 4**	28**
Either side of abdomen	6 & 2	12 & 4	32: 8*	24*
Midline of abdomen	6 & 1	12 & 2	32: 4**	28**

* Should be $32:10\frac{2}{3}$ and $21\frac{1}{3}$

** Should be $32:5\frac{1}{3}$ and $26\frac{2}{3}$

*** The text has simply *in chastrochemiis*, which is not a part of the body. It presumably refers to the desensitization of the heel.

The various parts of the human body possessing the touch-sense differ in sensitivity, and thus in their ability to distinguish weights. We have up till now estimated their sensitivity by the following method: we placed two different weights on the parts (hands, feet, shoulders, and other places), and observed the difference in weight necessary for clear tactile perception in a given part. A weight difference that is clearly perceived in the hands is, in fact, not discriminated on the shoulders, as we mentioned above. The shoulders are less sensitive than the hands, and so I increased the weight difference on the shoulders until it was just recognized. The heavier the weights that had to be added to the other shoulder to make the difference perceptible, the blunter the touch-sense.

The other method must also be used: different weights must be placed *simultaneously* rather than *sequentially* on the organs whose sensitivity we wish to compare and estimate, and a note made of the difference in weight required for the impressions we feel equally strong in these organs.

If equal weights are placed on various organs - one on the lips, for example, and another on the forehead - we do not feel equal pressure on both organs. [98] A weight pressing on the forehead seems less than the same weight on the lips, because the forehead has poorer tactile sensitivity than the lips. Now if we increase the weight on the forehead until the pressure seems equal on both places, the difference in weight gives numerical expression to the difference in sensitivity between the two places.

It is now appropriate to describe the experiments that I performed for that purpose.

(1) *A comparison of tactile sensitivity for weight estimation in the lips and forehead.*

In these experiments the head was held in a horizontal position, so that the weights could be placed on both parts with the head immobile and the muscles relaxed. A thick card was placed between the weights and the organs, to prevent their cold temperature from disturbing the judgement. A colleague placed the weights on my mouth and forehead, and changed them round repeatedly without my knowledge. I myself considered very carefully which weight felt heavier.

Number of Experiment	Weights placed on immobile lips	Weights placed on forehead	Sense of pressure
(1)	2 oz.	2 oz.	greater on lips
(2)	2	4	equal
(3)	1	4	equal
(4)	$\frac{1}{2}$	4	greater on forehead

From these experiments it can be concluded that tactile sensitivity for weight estimation is about *three times greater* in the *lips* than the *forehead*.

On a different occasion, several months earlier, another series of experiments was carried out on the same problem. However, I had forgotten about these when I carried out the second series.

Number of Experiment	Weights placed on lips	Weights placed on forehead	Sense of pressure
(1)	4 oz.	4 oz.	greater on lips
(2)	$1\frac{1}{2}$	4	greater on lips
(3)	1	4	equal

[99] From these experiments it can be concluded that tactile sensitivity in the *lips* is *three* or *four* times greater than in the forehead.

If we wish to compare tactile sensitivity in many places, it would be an excessively long and tedious undertaking to compare the individual parts with each other. It is adequate to select one part, such as the volar surface of the third and second phalanx of the fingers of the right hand, and compare them with other parts of the body.

That was the procedure I used, taking care that the transverse furrow separating the volar surface of the second and third phalanx touched the middle of the weight.

(2) *A comparison of tactile sensitivity for weight estimation in the volar surface of the fingers and in the lips*

Number of Experiment	Weights on volar surface of 2nd & 3rd phalanx of 3rd & 4th finger of right hand	Weights on lips	Sense of pressure
(1)	4 oz.	4 oz.	greater on lips
(2)	5	4	" " "
(3)	$5\frac{1}{2}$	4	" " "
(4)	6	4	" " "
(5)	$6\frac{1}{2}$	4	perhaps greater on lips
(6)	7	4	" " "
(7)	$7\frac{1}{2}$	4	" " "
(8)	8	4	equal pressure
(9)	9	4	" "
(10)	10	4	" "
(11)	11	4	perhaps greater on fingers
(12)	$11\frac{1}{2}$	4	greater on fingers

[100] From these experiments it appears that the pressure of the weights feels equal when 9 oz. was placed on the fingers and 4 oz. on the lips (exp. 9). It follows from this that tactile sensitivity in the lips bore the ratio of 1:2¼ to that in the fingers. For, if the fingers and the lips had been equally sensitive, 4 oz. would have produced an equal sensation of pressure in both places. In fact, 5 oz. (or 1¼ parts) was added to the 4 oz. (or 1 part) which had been placed on the fingers, to make the pressure equal in both places. Tactile sensitivity in the fingers was therefore 2¼ times less than in the lips. This conclusion is also confirmed in the following way: if 2½ oz. is subtracted (see exp. 5) from the 9 oz. that had been placed on the fingers in exp. 9, the pressure on the lips certainly feels greater than on the fingers; and conversely, if 2½ oz. are added (see exp. 12) to the same 9 oz., the pressure on the fingers certainly feels greater than on the lips. The 9 oz. mentioned in exp. 9 is the number midway between these.

If the weights could have been placed on the fingertips, it would certainly have been found that tactile sensitivity was better there than on the volar surface of the first and second phalanx.

However, in these experiments the weight lay over both phalanges simultaneously.

(3) *A comparison of tactile sensitivity on the volar surface of the fingers and on the forehead.*

The head and hand were held in a horizontal position, so that they were kept stationary without muscular action.

Number of experiment [101]	Weights on 2nd & 3rd phalanx of 3rd & 4th finger of right hand	Weights on mid forehead	Sense of pressure
(1)	4 oz.	4½	Certainly greater on forehead
(2)	4	4	Greater on forehead
(3)	4½	4	Perhaps greater on forehead
(4)	5	4	Equal pressure
(5)	5½	4	" "
(6)	6	4	" "
(7)	7	4	Perhaps greater on fingers
(8)	7½	4	" "
(9)	8	4	Much greater on fingers

From these experiments it appears that the pressure of the weights felt equal if (as I observed in exp. 5) about $5\frac{1}{2}$ oz. were placed on the fingers and 4 on the forehead; and the pressure felt greater in the fingers if (as in exp. 9) 8 oz. were placed on the fingers and 4 on the forehead; and, finally, it felt greater on the forehead if (as in exp. 1) $4\frac{1}{2}$ oz. were placed on the forehead and 4 oz. on the fingers. It can be concluded, then, that tactile sensitivity in that part of the fingers bore a ratio of 11:8 to that in the forehead.

(4) *A comparison of tactile sensitivity on the volar surface of the fingers and on the back of the head.*

Number of Experiments	Weights on volar surface of 2nd & 3rd phalanx of 3rd & 4th finger of right hand	Weights on middle of hairy back of head	Sense of pressure
(1)	8 oz.	4 oz.	Much great on fingers
(2)	$7\frac{1}{2}$	4	Greater on fingers
(3)	7	4	Greater on fingers
(4) [102]	6	4	Perhaps greater on fingers
(5)	5	4	" " "
(6)	4	4	" " "
(7)	4	5	Equal pressure
(8)	4	6	No clear difference
(9)	4	7	Perhaps greater on head
(10)	4	$7\frac{1}{2}$	Certainly greater on head

From these experiments it appears that the pressure of the weights seemed equal if (as in exp. 7) 4 oz. were placed on the volar surface of the fingers, and 5 oz. on the hairy part of the mid-occiput. The pressure seemed greater on the fingers if (as in exp. 3) 7 oz. were placed on the fingers and 4 oz. on the hairy occiput. Finally, the pressure was greater on the occiput if (as in exp. 10) 4 oz. were placed on the fingers and $7\frac{1}{2}$ on the occiput. It follows that the volar surface of the second and third phalanx of the finger is slightly superior to the occiput in tactile sensitivity, at least with regard to weight perception.

(5) *A comparison of tactile sensitivity on the volar surface of the fingers and the middle of the forearm*

The experiments were all performed on the same arm. In this way I avoided the errors otherwise arising from the different sensitivity of the right and left sides. However, it should be mentioned that tactile sensitivity for weight discrimination is greater if the weights are placed on different arms rather than the same arms.

Number of experiment [103]	Weights on volar surface of 2nd & 3rd phalanx of 3rd & 4th finger of right hand	Weights on volar surface of mid-forearm	Sense of pressure
(1)	$5\frac{1}{2}$ oz.	4 oz.	Greater on fingers
(2)	5	4	" " "
(3)	4	4	Perhaps greater on fingers
(4)	$4\frac{1}{2}$	4	Difference not clear
(5)	5	4	Perhaps greater on fingers
(6)	$5\frac{1}{2}$	4	" "
(7)	6	4	Greater on fingers
(8)	4	4	Difference not clear
(9)	4	5	" "
(10)	4	6	" "
(11)	4	$6\frac{1}{2}$	" "
(12)	4	7	Greater on forearm

From these experiments it follows that the pressure of the weights seemed equal if (as in exp. 9) 4 oz. were placed on the fingers and 5 oz. on the forearm; it would have seemed greater on the fingers if (as in exp. 2) 5 oz. were placed on the fingers; and, finally, it would have seemed greater on the forearm if (as in exp. 12) 7 oz. were placed on the forearm and 4 oz. on the fingers. The ratio of weights used in exp. 7, in which no weight difference was perceived*, is midway between those ratios for which the fingers (in exp. 2) and the forearm (in exp. 12) seemed to be pressed more strongly. It can be concluded from all these tests that tactile sensitivity in the fingers and the forearm bears the ratio of 5:4.

* This statement is inconsistent with the table. Weber probably intended to refer to exp. 9, which was indifferent and had a ratio of 4:5.

(6) *Tactile sensitivity on the volar surface of the fingers and the middle of the upper arm, or humerus.*

The experiments were carried out on one arm only. I kept the arm supported, as well as the hands and fingers.

Number of Experiment [104]	Weights on volar surface of 2nd & 3rd phalanx of 3rd & 4th finger	Weights on volar surface of mid upper arm	Sense of pressure
(1)	$4\frac{1}{2}$	4 oz.	Greater on fingers
(2)	4	4	" " "
(3)	4	$4\frac{1}{2}$	" " "
(4)	4	5	Equal pressure
(5)	4	$5\frac{1}{2}$	" "
(6)	4	6	Uncertain difference
(7)	4	$6\frac{1}{2}$	" "
(8)	4	7	Slightly greater *
(9)	4	$7\frac{1}{2}$	Greater on arm

The ratio of the sensitivity of the fingers to that of the arm is therefore $5\frac{3}{4}:4$.

The weights used in exp. 6 (and indeed also in exp. 5) are almost midway between those of exps. 3 and 9. That is why I have chosen these experiments to determine the ratio of the weights producing an equal sense of pressure on the fingers and arm.

The reason why I am comparing exps. 3 and 9 with exp. 6 is this: exp. 3 is the last in which the pressure of the weights is greater on the fingers, and exp. 9 is the first in which greater pressure is felt on the arm. Now if these experiments are compared with the sixth, it is obvious that the weights placed on the arm in exp. 6 differ by $2\frac{1}{2}$ oz. from those in exp. 3; however, the weights in exp. 6 differ by a greater amount (2 oz.) from those in exp. 9**.

We are sometimes mistaken in judging weights as equal, when they are placed simultaneously on the fingers and arm. In this type of experiment, then, it is strongly recommended that one should note the ratio of the weights, both when the pressure feels greater on the arm and when it feels greater on the fingers, in addition to noting the difference in weight when the pressure seems equal on the fingers and arm. [105] We can then infer

* Presumably 'on the arm' should be understood.

** It is not clear what point Weber is making here. The differences mentioned appear, from the table, to be $1\frac{1}{2}$ oz. in both cases.

the true mean from the three observations. In fact, if we wish to determine by sensation the ratio of the weights when the pressure seems definitely equal, we are much less certain and confident than if we determine the ratio when one weight feels slightly heavier in one or the other place. It is recommended, therefore, that for the estimation of tactile sensitivity we should examine the trials in which a difference was first perceived, rather than those where the pressure felt equal.

(7) *A comparison of tactile sensitivity on the volar surface of the fingers and the fold opposite the elbow.*

I held out the right arm, while lying supine on the floor, so that it was possible to place the weights on both the above places, while they were supported and completely still.

Number of experiment	Weights on fingers	Weights on fold of arm	Sense of pressure
(1)	$4\frac{1}{2}$	4 oz.	Greater on fingers
(2)	4	4	" " "
(3)	4	5	Difference uncertain
(4)	4	6	Perhaps greater on arm
(5)	4	$6\frac{1}{2}$	Difference uncertain
(6)	4	7	Greater pressure on arm
(7)	4	$7\frac{1}{2}$	" " "

From these experiments it can be concluded that the ratio of tactile sensitivity in the fingers to that in the fold of the arm is about 11:8. Tactile sensitivity in the fingers and mid-forearm had the ratio of 5:4, and in the fingers and mid-upper arm $5\frac{3}{4}$:4. It is clear, then, that the touch-sense in the fold of the arm has about the same degree of sensitivity as in the other two places, holding a rank almost midway between them. [106] It should be noted, however, that the median nerve in the fold of the arm is situated so that any weights placed on the fold must certainly cover and press it, except when it is protected by the skin and aponeurotic membrane of the biceps muscle. You might therefore expect the sense of pressure to be greater there, and weight discrimination more precise, than in surrounding areas. However, this was disproved by my experiments.

(8) *A comparison of tactile sensitivity on the volar surface of the fingers and on the internal side of the head of the brachial bone.*

I held out my right arm, while lying supine on the floor, so that it was possible to put the weights on the two places, while they were supported and immobile.

Number of experiment	Weights on fingers	Weights on head of brachial bone	Sense of pressure
(1)	5 oz.	8 oz.	Greater on fingers
(2)	4½	8	Slightly greater on fingers
(3)	4	8	Greater on fingers
(4)	3½	8	Equal pressure
(5)	3	8	Slightly greater on arm
(6)	2½	8	Greater on arm

These experiments show that tactile sensitivity for the fingers and for the skin on the internal side of the head of the brachial bone is almost in the ratio of 8:3½. For between the ratio 4½:8 (see exp. 2) and 2½:8 (see exp. 6), the geometric mean proportion is 6.71:16, which scarcely differs from the proportion that I stated.

(9) *A comparison of tactile sensitivity on the volar surface of the fingers and on the dorsal surface of the shoulder.*

I held out my right arm, while lying prone on the floor, so that it was possible to put the weights on the two places while they were supported and immobile.

Number of Experiment [107]	Weights on fingers	Weights on dorsal surface of shoulder	Sense of pressure
(1)	6 oz.	8 oz.	Greater on fingers
(2)	5	8	" " "
(3)	4½	8	Equal pressure
(4)	4	8	Perhaps greater on fingers
(5)	3½	8	Perhaps greater on shoulder
(6)	3	8	Difference uncertain
(7)	2½	8	" "
(8)	2	8	Perhaps greater on shoulder
(9)	1½	8	" " "
(10)	1	8	Greater on shoulder

These experiments show that tactile sensitivity for the fingers and for the dorsal surface of the shoulder was almost in the ratio of 16:5½*, if we take the geometric mean between the proportions 5:8 (see exp. 2) and 1:8 (see exp. 10).

(10) *A comparison of tactile sensitivity on the volar surface of the fingers and on the plantar surface of the big toe.*

I lay prone on the floor and spread the toes of my arched right foot on a wooden stool. I also held out my arm, so that it was possible to put the weights on both places while they were supported and immobile.

Number of experiment	Weights on fingers	Weights on plantar surface of toe	Sense of pressure
(1)	4 oz.	6 oz.	Greater on fingers
(2)	4	6½	Slightly greater on fingers
(3)	4	7	Equal pressure
(4)	4	8	" "
(5)	4	9	" "
(6)	4	10	Perhaps greater on toe
(7)	4	11	Greater on toe

[108] The ratio of tactile sensitivity for the fingers and the plantar surface of the toe can be derived from the above table. This ratio is approximately 2:1; for between the ratios 4:6 (see exp. 1) and 4:10 (see exp. 6) the geometric mean proportion is 8:16.25**.

(11) *A comparison of tactile sensitivity on the volar surface of the fingers and on the metatarsal capitulum of the big toe.*

The position of the body was the same as in the previous experiment.

* An inaccurate representation of 4.47:16

** An inaccurate representation of 8:15.5

Number of experiment	Weights on fingers	Weights on plantar surface of metatarsal capitulum of big toe	Sense of pressure
(1)	4 oz.	9 oz.	Greater on fingers
(2)	4	9 ½	Equal pressure
(3)	4	10 ½	" "
(4)	4	11	Perhaps greater on foot
(5)	4	12	Greater on foot.

It is clear from these experiments that tactile sensitivity in the fingers and in the metatarsal capitulum of the big toe corresponds approximately to the ratio 20¾:8. For between the ratios 4:9 (see exp. 1) and 4:12 (see exp. 5) you will find that the geometric mean proportion is 8:20.78.

(12) *A comparison of tactile sensitivity on the volar surface of the fingers and on the heel.*

The position of the body was the same as in the previous experiment.

Number of experiment	Weights on volar surface of fingers	Weights on heel	Sense of pressure
(1)	4 oz.	5½	Greater pressure on fingers
(2)	4	6½	Perhaps greater on fingers
(3)	4	7½	" " "
(4)	4	8	Equal pressure
(5)	4	9	" "
(6) [109]	4	10	" "
(7)	4	10½	Difference uncertain
(8)	4	11	Perhaps greater on heel
(9)	4	11½	Greater on heel

Tactile sensitivity for the fingers and the heel can therefore be expressed by the numbers 16:8. For if you calculate the geometric mean proportion between the ratios 4:5½ (see exp. 1) and 4:11½ (see exp. 9), you will get 8:15.91.

(13) *A comparison of tactile sensitivity on the volar surface of the fingers and on the concave part of the sole of the foot.*

Number of experiment	Weights on fingers	Weights on sole	Sense of pressure
(1)	4 oz.	5 oz.	Greater on fingers
(2)	4	6	Slightly greater on fingers
(3)	4	7	Perhaps greater on fingers
(4)	4	$7\frac{1}{2}$	" " "
(5)	4	9	Equal pressure
(6)	4	10	" "
(7)	4	$12\frac{1}{2}$	Difference uncertain
(8)	4	13	Perhaps greater on sole
(9)	4	14	" " "
(10)	4	15	Slightly greater on sole
(11)	4	$15\frac{1}{2}$	Greater on sole

If you do the calculations from these experiments, you will find that tactile sensitivity in the fingers and the concave part of the sole has approximately the ratio of $18\frac{1}{2}$:8. For the geometric mean between the ratios 4:$15\frac{1}{2}$ and 4:5 corresponds to the ratio 8:18.47*.

(14) *A comparison of tactile sensitivity on the volar surface of the fingers and behind the knee.* [110]

I lay prone on the floor, with the right foot slightly flexed and well supported.

Number of experiment	Weights on fingers	Weights behind knee	Sense of pressure
(1)	4 oz.	$4\frac{1}{2}$	Greater on fingers
(2)	4	5	Perhaps greater on fingers
(3)	4	$5\frac{1}{2}$	Equal pressure
(4)	4	6	" "
(5)	4	7	" "
(6)	4	8	Perhaps greater on leg
(7)	4	9	" " "
(8)	4	$9\frac{1}{2}$	Greater on back of knee

* An inaccurate representation of 8:17.6.

By the same rule as we have used previously, tactile sensitivity for the fingers and the back of the knee is almost equal in ratio to the numbers 13:8. The geometric mean between the ratios 8:4½ (see exp. 1) and 4:9½ (see exp. 8) is in fact 8:13.08.

(15) *A comparison of sensitivity on the volar surface of the fingers and on the lateral part of the knee near the internal condyle.*

Number of experiment	Weights on fingers	Weights on internal side of knee	Sense of pressure
(1)	4 oz.	11½	Greater on fingers
(2)	4	12	Difference uncertain
(3)	4	13	" "
(4)	4	13½	Greater on knee

I conclude from these data that tactile sensitivity in the fingers and the lateral part of the knee corresponds to the numbers 25:8, while I calculate the geometric mean between the ratios 4:13½ (see exp. 4) and 4:11½ as 8:24.92.

[111]Now that these experiments have been explained, let us see what should be repeated to illustrate the variation in tactile sensitivity over different parts of the skin. I might compare our weight-sensitive organs with the arms of weighing-scales of unequal length. If the arms are unequal, and equal weights are placed on them, the weight on the longer arm has a greater effect than the one on the shorter arm. Different weights must be used in order to restore equilibrium to the scales and produce an equal effect on the two arms. The difference in length between the arms is easily calculated, if we know the difference between the weights necessary to restore equilibrium.

Something similar to the weighing-scales happens to us, if weights are placed simultaneously on different parts of the body with unequal tactile sensitivity. Equal weights do not produce a sense of equilibrium; instead, a much heavier weight is often needed on the less sensitive part before we seem to feel equal pressure in the two places.

We use the same method of calculation for the touch-organs as for the arms of the scales: we measure tactile sensitivity in various parts of the body from the different weights necessary to restore equilibrium.

As a matter of fact, all the touch-organs are at best very insensitive. For example, if you place two weights simultaneously on the same hand, one on the metacarpal capitulum of the index finger and the other on the same part of the little finger, you will not feel the difference in weight unless it is large. Sensitivity* varies in different places, and I felt it useful to measure the variation by comparing all places with the fingers. We intended then, by the experiments described earlier, to find the difference in sensitivity in different places. Some immobility also occurs in the two arms of weighing scales, and is the reason why very small weight differences on the arms are not very clearly reregistered. [112] Physicists call it the inertia** of the scales. However, this immobility is minimal in well-constructed scales, and even in poorly-constructed scales it is much less than the sluggishness*** of the touch-organs.

Anyhow, this immobility of the scales is not related to the fact that the longer arm is pressed down more than the shorter arm by an equal weight. The immobility has other causes, such as the location of the point of equilibrium in the beam of the scales, and on the distance between that and the point of suspension of the beam. It also depends partly on the sharpness of the prism, which is more or less blunt: when it is interposed, the beam lies on a flat surface, and sticks to it.****

The degree of sluggishness that is shared by any two touch-organs should, in my opinion, be compared with the inertia of the scales which impedes the responses of the two arms to exactly the same extent. However, I feel justified in comparing the difference in sluggishness between two organs with the effect of a difference in length between the two arms of the scales. For, a great degree of sluggishness and a shortness of one or other arm of the scales have the same result, that a given weight has less effect on one side of the body or of the scales than on the other.

---

* Weber uses the negative word *torpor* (sluggishness, or insensitivity) throughout this passage, and compares the *torpor* of the touch-organs with the inertia of weighing scales. This negative description of sensitivity is difficult to bring out in translation. *Subtilitas* or *sensibilitas* are the positive words that Weber normally uses for sensitivity. He also uses *subtilitas* to refer to tactile acuity for the two-point threshold; but in this analogy with weighing scales he is working his way towards a distinction between acuity and sensitivity.

** *pigritiem*

*** *torpor*

**** *partim ab acie prismatis plus minusve obtusa, qua interposita vectis tabulae planae incumbit, eique adhaeret.* Weber refers to a steelyard, or balance with unequal arms, whose fulcrum is a prism-shaped knife-edge.

The following proposition about the strength of irritation of the nerves is amply illustrated and confirmed by my experiments: the strength of nervous irritation certainly depends on two causes - on the strength of the irritating stimulus and on the degree of irritability of the irritated part*. The irritability of a part increases when it is supplied with a greater number of nerves, and when the brain and nerves are better designed for receiving impressions.

If a certain part is less irritable, the same irritating stimulus will arouse a weaker sensation than in a more irritable part. When the irritating stimulus is increased to some extent, the strength of the sensation is also increased; but there is no increase in the difference in sensation between the less irritable and more irritable part.

The sluggishness that is common to two touch-organs can also be elucidated from the tests related above. We can discover it by calculating the number of ounces that could be added to or subtracted from the weights on the touch-organs without a change in the sense of pressure.[113]

It must be mentioned that in no type of experiment were the observations sufficiently certain and constant for it to be possible to determine the sluggishness of the touch-sense of any individual organs. We must therefore accept the fact that our conclusions on this subject are of a general nature.

We have now discussed the way in which we perceive the separation and the weight of objects, when our touch-organs are stationary or moving. We must next bring out a third property of the touch-organ.

ON TACTILE SENSITIVITY IN THE PERCEPTION OF TEMPERATURE

We have still to mention the use of the touch-organs for estimating the temperature of objects. 'Estimation' is the word I use rather than 'perception', in relation to temperature; for we perceive the effect of an increase or decrease in heat produced in the touch-organ, and from that we judge the temperature of the objects. We should often make errors in estimating the temperature of objects, if we considered only the sensation of cold or heat perceived by the touch-organ. Some objects - such as metals, for example - quickly lose a large part of their heat when they come into contact with colder objects. If such objects touch our skin, our first naive impression is that they are warmer than other objects of the same temperature which - like wood - lose heat more slowly. The

* *vehementium nimirum irritationis nervorum a duabus causis pendere, a vehementia irritamenti et a gradu irritabilitatis partis irritatae.*

same thing applies to the sense of cold, and is easily explained. The greater the amount of heat withdrawn from some object by our skin in a certain time, and the greater the amount of heat that consequently accumulates in our skin, the stronger the sensation of heat; and conversely, the greater the amount of heat withdrawn from our skin, the stronger the sensation of cold.

However, this is not the only source of error in temperature estimation. Another error arises with warm objects, because they lose more heat when they cover a larger area of skin. It is easy for anyone to test this, if he fills two flasks with warm water and then asks a subject to put his hand in one flask but only one finger in the other. [114] The water will feel warmer when it surrounds the whole hand than when it is in contact with only one finger. The subject's sensation of heat is increased by the large area of the immersed hand, and diminished by the very small circumference of the finger. The difference is so great that water feels warmer to the hand than to the finger, even if it is colder by two or more degrees. When carrying out this experiment, I filled two wooden flasks with water that was sufficiently warm to arouse a sensation of heat in the skin. Réaumurian* thermometers placed in the flasks showed a temperature of 29½° in one flask and 33° in the other. If a colleague who did not know the water temperature immersed his finger in the warmer water and simultaneously put his whole hand in the colder water, it was the latter that felt warmer. Another colleague also tried the same experiment when the water in one flask was 29½° and the other 32°. In both tests the heat sensation aroused in the hand was greater than that in the finger - but only at first, when the difference in temperature was unequivocal. However, if the hand and finger were kept immersed in the water for a long time, the first impetus of the stimulation subsided and the judgement of temperature was often more accurate.

Water at a temperature of 19° or 17° arouses a sensation of cold rather than of heat in the fingers and hand. If you perform a similar experiment at this temperature, you will find the same for the sense of cold as has just been reported for the sense of heat: the whole hand is affected by a stronger sensation of cold than the finger, even when immersed in less cold water.

This effect clearly depends on the fact that the impressions made by single particles of water in different points of the skin merge together in one impression:

---

* The Réaumur temperature scale was established in 1730 by the French naturalist René-Antoine Ferchault de Réaumur (1683-1757). The scale has zero set at the freezing point of water, and 80° at the boiling point of water at normal atmospheric pressure. 1°R = 1.25°C.

therefore a *smaller degree* of heat received by a *larger surface* of the skin arouses the same sensation as a *greater degree* of heat stimulating a *smaller surface* of the skin.

Other experiments also demonstrate this merging together of heat impressions.

[115] This summation of sensations obscures our discrimination. We are less able to distinguish the different temperatures of water in two adjacent flasks, if we put the thumb in one flask and the second finger on the same hand in the other flask at the same time. The difference is distinguished more easily and accurately if we immerse both thumbs in the flasks simultaneously: for the impressions from touch-organs that are used simultaneously merge less easily if they are further apart. In fact, the best and most accurate tactile discrimination of temperature is made if the same finger, or different fingers, are immersed successively rather than simultaneously.

It is the merging of impressions, and not the distraction of the mind when disturbed by two simultaneous impressions, that hinders temperature discrimination. This is clear, because if the observation is continued for even longer and the attention is turned alternately to the thumb and the second finger, the temperature still cannot be discriminated any better. If the two sense impressions did not merge together simultaneously, it should be possible for us to obscure one by turning the attention towards it.

The lengths of two lines drawn on paper can be discriminated more precisely if the eye takes them both in simultaneously rather than inspecting them successively. In the latter case the line inspected first must be committed to memory, and its image reproduced by the power of the imagination* for comparison with the next line. This very exceptional power of memory and imagination is important, because it enables us to make a very accurate comparison of successive heat impressions on the touch-organ, even if some time elapses between each sensation.

If I were asked whether a similar merging of two sensations also occurs in the eye, I would not completely deny it. However, such a merging arises rarely in vision, and only for specific reasons. For different colours are perceived as juxtaposed, [116] and they merge together only if they are a great distance from the eyes and very close to each other. In fact they never merge completely, but only at the edge where one colour touches another.

This is relevant to an experiment first performed by Du Tour**, in which one looks through differently colour-

---

* *phantasiae vi*

** Étienne François Du Tour (1711-1784) was a French physicist who worked on electricity and optics (especially the diffraction of light). Among his papers was "Pourquoi un objet sur lequel nous fixons lex yeux, paroit-il unique?" *Mém.Par. sav. étrang.* Vol. III, 1760.

ed pieces of glass held close to the eyes, so that the long axes of the eyes converge on the object of inspection. It must be admitted that not all physiologists repeating this experiment have described the same effect. Janin* and many others contend that the impression of the colours of both pieces of glass merge into one, and form an intermediate colour. Johannes Müller** was the most recent writer on this subject (*Zur vergleichenden Physiologie des Gesichtssinnes, Leipzig* 1826, 8, p.80); he tried many experiments and opposed this conclusion. He asserted that the intermediate colour did not appear, but that the reduction in brightness caused by the coloured pieces of glass is lessened if one piece has a clearer and more brilliant colour[8], and that the objects appear to alternate between the two colours.

The view of Müller agrees with the observation of these earlier physiologists in so far as he asserts that an object seen through two differently coloured glasses appears to have one colour rather than two; but it differs in that he denies that the object appears to be covered by a third colour, and he asserts on the contrary that it appears to be covered with each of the different colours in turn.

I myself was not led towards Müller's position by the tests carried out on myself and others.

In fact, if the eyes are turned so that the long axes converge on the object, and if they are almost equally sound as regards short or long sightedness, the objects always appear to be covered with one colour. [117] This differs in another way besides the brightness***from the apparent colour that they seemed to possess through single pieces of glass.

I admit that I have not definitely seen a green colour when I placed yellow glass in front of one eye and blue in front of the other. Nevertheless, the combination of the yellow and blue glass almost completely removed the predominant cast produced by one colour alone. When the original colours were completely gone, another colour would usually replace them - one that was very desaturated, impure, mixed, and difficult to recognize as a definite colour. However, when the original colours did not disappear completely, the colour of the more strongly coloured glass would usually dominate, though with a very desaturated appearance.

---

* Probably Jean Janin de Combe Blanche, (c. 1731-1811), an oculist, and Honorary Professor at the University of Modena. He was the author of *Memoires et observations anatomiques, physiologiques et physiques sur l'oeil*, Lyon, 1772.

** Johannes Müller (1801-1858) was Professor of Anatomy and Physiology at Berlin from 1833. He is best known for his doctrine of specific nerve energies.

*** *caligo*, darkness.

I maintain, nevertheless, that the impressions of the colours merge into one if we look through differently coloured pieces of glass placed simultaneously in front of the two eyes. It is quite certain that objects are perceived as single rather than double with binocular vision, provided that both eyes are normal and equally strong. Objects are also perceived by the two eyes simultaneously rather than alternately; and they appear to have a different colour from the one seen through either coloured glass alone. I therefore deny that objects seem to alternate between a blue and yellow colour when both colours of glass are used together - provided that the eyes are normal and equally strong and are focussed so that their long axes converge on the test object. It is up to physicists to explain why objects do not appear to have a green colour. The same thing is observed if light is mixed in a particular way while passing through the same pieces of glass. If you set up a perpendicular sheet of yellow glass in a room with two windows, so that the light entering the room through each window penetrates the glass and falls on a sheet of white paper, you will find that this light is a translucent yellow. If you place blue glass opposite the window in the same way, you will notice that the penetrating light is blue. However, if you place both sheets of glass beside each other so that part of the light transmitted through each falls on the same part of the paper, and so is mixed, you will see that no green colour occurs. In my experiments the yellow and blue colours tended slightly towards the red, and the rays of light that were transmitted and were intermingled on the paper formed a colour commonly called violet. [118] However, if the yellow glass was placed in front of the blue, the light penetrating through both sheets and falling on the paper possessed a colour that was predominantly green. This shows that it is possible to amalgamate coloured rays so that the resultant colour is not the one that we usually find as the intermediate colour*.

Anyone wishing to repeat the experiments (concerning the coloured pieces of glass placed in front of both eyes simultaneously) should take care that the axes of both eyes converge on the same spot. When first looking through the pieces of coloured glass, one easily directs the axes of the eyes so that the right eye inspects one part of the scene and the left eye another. In this case any part appears tinted with the colour of the glass in front of the relevant eye. Observers should also attend

---

* Weber is describing the difference between additive and subtractive colour mixtures. He assumes that subtractive colour mixture (in which yellow and blue broad-band filters produce green) is normal, and that additive colour mixture is in need of explanation. At this time there was still widespread confusion as to whether the perceived effects of colour mixture were a property of man or a property of the world.

to objects in the middle of the visual scene, and ignore things on the periphery. Next, the object should be placed at a suitable distance from the eyes, so that a single and clear image is formed. I perceived a candle flame as a single image with a definite outline only when I moved it sufficiently close to the eyes and the pieces of glass. If we see distant objects as single and possessing one colour, nearer objects often appear double and covered with the colours of the two pieces of glass. Subjects whose eyes are unequal in strength usually find that the colour of the glass in front of the strong eye prevails. Subjects who are used to attending to each eye alternately while looking through the coloured pieces of glass will see objects taking on the colour of each glass in turn; and indeed they often notice differently coloured spots on the objects. Finally, it should be mentioned that the experiments are more easily carried out if the two pieces of glass are tinted with equally strong colours, or if the glass with the paler colour is thicker.

It follows from all these reports that the sensation of two colours, when one is seen by the right eye and the other by the left, never merges into one except when the colours seem to come from the same direction. [119] This happens only if the images of the colours occupy exactly the same place in the two eyes. In all other cases we seem to distinguish several colours, which are either seen by the same eye, or are perceived so that one colour is seen by the right eye and another by the left.

I come now to a remarkable observation about another error to which we are exposed when judging temperature:* *liquids of equal temperature do not affect the left hand in the same way as the right, but for most subjects a stronger sensation of heat or cold arises in the left hand*. This difference between the two sides is worth noting, especially since I have shown in other experiments that equal weights placed simultaneously on both hands arouse, in most subjects, a stronger sensation of pressure in the left hand than in the right.

First I used two accurate thermometers to examine the temperature of the hands in a number of subjects. Their hands had been covered in the same manner for a long time, or had been exposed to the same general conditions. The subjects then took hold of the two thermometers, totally enclosing them. In this way it was found that immediately after the beginning of the experiment the mercury in the left hand rose a half or a whole degree higher than in the right. This difference in tem-

---

* An alternative translation of the following section [pp. 119-122] by J.D. Mollon appears in Kinsbourne (op. cit. p.57)

perature gradually declined the longer the experiment was continued, so that at the end of a long experiment the temperature in both hands was equal or almost equal (i.e. the difference was a third or a quarter of a degree). The hands were found to have a temperature of $28\frac{1}{2}^{\circ}$, 29°, $29\frac{1}{3}^{\circ}$ or $29\frac{2}{3}^{\circ}$ on the Réaumurian scale, though a slightly higher temperature was observed transiently, sometimes on the left and sometimes on the right. Next, I filled two large wooden flasks with warm water, and checked their temperature with two thermometers that were permanently immersed in the water. Then I asked a colleague, who did not know the temperature, to immerse both index fingers or hands in the water, at the same time and to the same depth, and to judge the temperature of the water. [120]

In this way, all the while monitoring the thermometers, I found the result mentioned above, that *most subjects felt the water examined by the left hand as warmer than that investigated by the right hand,* even though there was no difference in temperature. If the position of the flasks was changed round without my colleague's knowledge, then it was always the water in which the left hand was immersed that seemed the warmer.

The same result was also observed even if the water for the right hand was a half or a whole degree warmer than that for the left hand.

In order to illustrate these statements by examples, I now present some experimental results.

Part of body immersed in water	Temperature of water	Part of body immersed in water	Temperature of water	Part with warmer sensation
Left hand	$33\frac{1}{2}^{\circ}$	Right hand	34°R	Left hand
Left index finger	33°	Right index finger	34°	Left finger
Left little finger	$31\frac{1}{2}^{\circ}$	Right little finger	32°	Left finger, but uncertainly
Left hand	31°	Right hand	$32\frac{1}{3}^{\circ}$	Neither hand
Left index finger	29°	Right index finger	31°	Neither finger
Left hand	$29\frac{1}{2}^{\circ}$	Right hand	$29\frac{1}{2}^{\circ}$	Left hand

If the water has a temperature of 19° or less, it arouses in the hand a sensation of cold rather than warmth. At this temperature, if both hands are immersed simultaneously, the left receives a stronger sensation of cold than the right. So it follows that both the accumulation of heat, and also the lack of it, stimulates the left hand more strongly.

Part of body immersed in water	Temperature of water	Part of body immersed in water	Temperature of water	Part with warmer sensation
Left thumb & index finger	$10\frac{1}{2}^{o}$ R	Right thumb & index finger	$10\frac{1}{2}^{o}$ R	Left fingers
Same fingers	$11\frac{3}{4}^{o}$	Same fingers	$10\frac{2}{3}^{o}$	Right fingers
Same fingers	12°	Same fingers	12°	Left fingers
Same fingers	12°	Same fingers	$11\frac{1}{2}^{o}$	Left fingers

[121] Of course we should not expect it to be true of all subjects that the left hand is more sensitive to heat than the right; but my experiments show that it is so in most cases. As to the reason why the left hand is more sensitive to heat and cold, it might be supposed most attention should be paid to the fact mentioned above, that the left hand is usually warmer than the right.

However, the falsity of this explanation is shown by the fact that the same result is observed in both cold and warm water. If the left hand were more sensitive to cold water because of its higher temperature, then it should necessarily be less sensitive to warm water than the right; for the temperature of the water would differ less from that of the left hand than from that of the colder right hand. But this goes against the facts, since the left hand is more strongly affected than the right, whether you put the hands in warmer or colder water. So we must have recourse to another explanation. The palm of the right hand is covered with thicker skin than that of the left. Since the thicker skin is less easily penetrated by heat or cold, it seems to produce both effects. Firstly, if the bulb of a thermometer is enclosed in the right hand, a given quantity of heat does not flow as quickly into the thermometer as from the left hand. Secondly, if the hands are immersed, a given quantity of heat or cold does not pass as quickly from the hot or cold water into the right hand as into the left. This explanation is also supported by another phenomenon which I observed on the back and palm of the hands. A fairly large wooden container is filled with water at a temperature of $+9\frac{1}{2}^{o}$, and both hands are opened out and immersed together. The ulnar surface is turned downwards and the radial surface upwards, and the two hands are slightly separated from each other. At the same time I pay attention to whether the back of the hand or the palm is more strongly stimulated by the cold. I always find [122] that at first the water touching the back of the hands feels colder than the water between them, but after the elapse of 10 or 15 or 24 seconds the cold sensation gradually lessens in the back of the hands and increases in the palm, so that eventually the water between the two hands feels much colder than the water touching the backs. I notice the same thing if the water is not so cold. Indeed, if I use warm water, I find that at first the water touching the back of my hand feels warmer, but after a few seconds the palm seems more strongly stimulated by the heat. I have observed this in the case of a colleague as well as in my own case, when the water had a temperature of $38\frac{1}{2}^{\circ}$.

When I placed both hands in warm water of 22°, the back was initially affected by a slight sensation of warmth and the palm by one of cold. I think the following is a plausible explanation for these findings: since the back of the hand has a thinner skin than the palm, heat and cold stimulate the nerves of the back more easily than those of the palm, and so the sensation of heat or cold arising from the water is initially stronger in the back. Because the palm also contains many more nerves and so has the finer sensitivity, the heat and cold gradually come to have a stronger and more consistent effect in the palm, once they have overcome the obstacle of the thicker skin.

Up till now we have been trying to bring out all the mistakes that physiologists make when investigating the sensitivity of the touch-sense as regards the estimation of temperature. We must now relate the conclusions about temperature sensitivity.

The sensitivity of the touch-sense is so much superior to my expectations that conventionally constructed thermometers are hardly of sufficient accuracy to measure the difference in temperature of the two liquids that subjects cannot quite perceive when they immerse their hands alternately in the liquids. Those subjects whom I examined could confidently distinguish a difference of a fifth of a degree on the Réaumurian scale; and it is not impossible that they could have distinguished a difference of a sixth or seventh of a degree. [123]

When I discovered such sensitivity, I wanted to know whether very small differences would be perceived more easily if the temperature of the liquids was so much warmer than that of the body that it only just failed to arouse a burning sensation. It is reasonable to suppose that we should feel the slightest changes of temperature, when objects are so hot that any increase in temperature would produce a burning sensation.

However, it is a fact that a difference between sounds is more easily heard if the sounds are neither too high nor too low, and if the intensity is so moderate that it does not cause annoyance. Similarly, the temperature is perceived most accurately when it does not hurt the fingers. This was clearly shown by experiments. Moreover, the temperature was recognized most accurately when it was very close to that of the hands, although a very small amount of heat was exchanged with them. It has in fact been shown above that the sensation of temperature is clearer and finer if a larger area of skin comes into contact with the liquid. So no one will be surprised that I did not feel the difference between two liquids of $20\frac{3}{4}$ and $20\frac{2}{3}^{\circ}$ R when I immersed only the index finger in the two liquids alternately, but that I distinguished them when I immersed the whole hand.

This research is considerably hindered if we use insufficient accuracy in the examination of tactile sensitivity in different parts of the body, and if we fail to compare the parts with each other.

Many people may think it is extraordinary that the parts that excel in tactile acuity (such as the tips of the fingers, the tip of the tongue, the lips, and other places) are not as good at temperature discrimination as they are at weight estimation or at two-point discrimination. For according to my experiments, the tongue and the finger-tips certainly have the best tactile acuity, but this advantage is insufficient to compensate for their small area in comparison with other parts. [124] For example, the following experiment was performed. The water in one flask had a temperature of $17\frac{3}{4}^{\circ}$R and the other 18°R, and the position of the two flasks was repeatedly interchanged without the knowledge of my colleague (the subject). My colleague tried to discriminate the difference in temperature, first by immersing the first phalanx of the index finger successively in the two flasks, then by immersing the lips, and finally by immersing the tip of the nose. The difference was always discriminated by this method, but the sensation was clearest in the finger, then the lips, and least clear in the nose. The same colleague was actually able to perceive the same difference in temperature with a flexed elbow - an area that has far lower tactile acuity than the fingers and lips. The difference previously discriminated by the first phalanx of the index finger was not less than $1/5^{\circ}$R, and the same colleague could not discriminate a smaller difference than 1/5 - 1/6 of a degree with the whole hand immersed. Therefore the individual parts of the skin seemed to differ not so much in the size of the difference threshold but rather in the clarity with which the same difference is recognized.

In these experiments it is difficult to arrange it so that the same area of skin touches the warm liquids for the elbow, lips, tongue and finger-tips; and yet the acuity and clarity of the sensation increases with the area of the skin. I therefore used another method to examine the acuity of the touch-sense for temperature perception. I filled two small phials with equal amounts of olive oil, and immersed two thermometers in them to measure the temperature of the oil. I sealed off the remaining part of the opening that was not filled by the thermometers, and placed the phials in hot water. I heated the oil very gradually, while the thermometers indicated its temperature. After noting the temperature, I took out the phials and dried them, and moved them to various parts of the skin. [125] In this way I discovered that the skin of the face generally, and especially that of the eye-lids and cheeks, was more sensitive to heat than the skin of other parts. The forehead, ears, lips, and other parts with very acute tactile sensitivity, are inferior to the eye-lids and cheeks in thermal sensitivity. I frequently moved the same phial to two places alternately, or moved it from one place to another and then removed it, and so tried to discover the varying effect of heat on the skin. Using this method, the middle part of the upper lip seemed less sensitive to heat than its sides. However, I realised that this method was also inadequate for judging the temperature sensitivity of different places. For the stronger effect of the heat seemed to depend initially on the thinner cuticle, which offered less resistance to the passage of heat. I confess that I am unable, for this reason, to use the method for estimating that type of tactile acuity, though the method is useful and clearly safe.

I come to the *coenaesthesis* of the skin - the sense by which we perceive pain if too much heat is given to or taken from the skin.

All perceptions, you see, are divided into two classes: the *objective* and the *subjective*. Objective perceptions arise when we not only sense the change in our organs caused by the stimulus of the perceived objects but also seem to sense the object itself stimulating our organs. For example, when light stimulates the nervous membrane of the eye, we do not perceive only the change in the eye: we also distinguish between the light and the objects producing the light - or so it seems. The same thing happens in the ear when we hear sounds. We also have the same experience with the tactile organs and the other organs, which are called the organs of sense for that very reason. For the unique property of the sense-organs is that they assist us to become the partakers of objective perceptions. The other parts of the human body lack this ability, and admit only *subjective* perceptions: their nature is such that we perceive only changes caused by our organs coming into contact with other objects. These perceptions are generally either pleasant or unpleasant, and the unpleasant ones are sometimes called 'pains'. For example, after we have cut a patient's skin with a knife, only the pain is perceived, and not the smoothness, temperature and shape of the skin of the adjacent cut parts. [126] Similarly, the muscles do not feel each other by some tactile sense during contractions; and the heart does not feel the ribs that it beats against.

I do not now wish to enquire into such an obscure

question as the structure responsible for this perceptual difference. Perhaps it lies in the structure of the organs and the ending of the nerves; or perhaps in the origin, course and nature of the nerves leading from the brain to the organs*. One point may be mentioned: the effect of diseases often destroys the ability of sense-organs to communicate *objective* perceptions to the mind, while the same organs remain able to arouse *subjective* perceptions. In this way I have seen severely blind patients suffering great pain, caused mainly by the light and various colours. It is also well known that hemiplegic patients, or those with some other paralysis of the skin, are usually unable to recognize heat, weight, and other aspects of objects by the touch-sense; but they experience severe pain in the skin if the limbs are rubbed, even in a gentle manner.

Moderate heat is recognized by objective perception: the heat touching and penetrating the organs is distinguished by the organs in some way, and its strength perceived. However, stronger heat arouses a pain called 'burning', in which case we do not perceive the heat but the changed condition of the organs. The same is also true of cold. Some people may suppose that these pains are caused by the extreme expansion or contraction of parts of the nerves; for it is known that heat expands and rarifies objects, while cold contracts and condenses them. Others believe that the pains are caused by the penetration of heat or cold to deeper places, especially to the trunks of the nerves. However, neither opinion seems to be favoured by the following experimental result: if the joint of the index finger, or of any other finger, is immersed for a long time in fairly hot water, it does not feel burned immediately; but several fingers immersed together do feel burned. Similarly, water that is not quite as hot does not burn the fingers, but does burn the whole hand submerged. It follows that it is unnecessary for individual particles of the nervous substance of the skin to expand much in order to arouse a burning sensation. [127] The expansion of the particles is not increased when you immerse the whole hand rather than just one finger; instead, a single finger heats up to the same degree as the whole hand. The burning sensation is caused because many particles of the skin become hot. If the individual particles each heat up to the same degree they will not provoke a burning sensation; but when they heat up together they produce a summated effect in the nerves - a burning sensation.

The sensation of burning depends on five factors: first, on the degree of heat that is transferred to the

---

* .. *utrum magis in organorum structura nervorumque fine, an in nervorum a cerebro ad organa deductorum origine, decursu et natura quaerenda sit.*

skin; then on the time for which the transfer continues; next, on the area of skin that comes into contact with the warmer object; then, on the thickness of the skin, thinner skin causing less impedance to the transfer of heat; and finally on the sensitivity of the part, since the parts vary in sensitivity.

Some of these factors may now be illustrated more fully. The time necessary to cause a burning sensation increases as the heat communicated to the skin is reduced, and as the area of skin touching the hot object is reduced. This follows from the next experiments.

A colleague placed the first joint of his index finger in hot water, the temperature of which I had measured, and indicated the exact moment when the sensation of burning became so intense that he was compelled to remove his finger. I measured the elapsed time on a clock with a second hand, and noticed that more time was required to excite a burning sensation the lower the water temperature*.

Temperature of water in which 1st joint of index finger immersed (in ^{o}R)		Time elapsed between immersion and retraction of finger (in seconds)
52°		$\frac{3}{4}$
51°		4
$50\frac{1}{3}^{o}$		$4\frac{1}{2}$
$49\frac{2}{3}^{o}$		$5\frac{1}{2}$
49°	[128]	$5\frac{1}{4}$
48°		7
$47\frac{2}{3}^{o}$		7
47°		9
$45\frac{3}{4}^{o}$		11
$45\frac{1}{2}^{o}$		$13\frac{1}{2}$
45°		14
$44\frac{1}{2}^{o}$		17
44°		21
$43\frac{1}{2}^{o}$		20
$42\frac{1}{2}^{o}$		23

If the water temperature was $41\frac{1}{2}^{o}$ the sensation of burning was certainly not aroused.

In these experiments care must be taken not to put a finger in the water again when it is still rather hot or irritated from its previous heating - for then, of course, the burning sensation would arise more quickly. For this reason, my colleague put his right and left index fingers alternately in the water.

* The table below is almost the same as the one on p. [134] of *Der Tastsinn*.

When I carried out these experiments on myself, I often thought it better to immerse my other smaller fingers in addition to the index finger. In this way I hoped to avoid depressing the sensitivity of the fingers by using them too often. These, then were the observations that I made*:

Temperature of water in which 1st joint of index finger immersed (in °R)	Time elapsed between immersion and retraction of finger (in seconds)
57°	$3\frac{1}{2}$
53°	$4\frac{1}{2}$
52°	4
51°	5
51°	4
50°	4
49°	8
48°	$5\frac{1}{2}$
56° [129]	$2\frac{1}{2}$
55°	$3\frac{1}{2}$
54°	$3\frac{1}{2}$
53°	4
52°	4
51°	5
50°	5

I noted above that the different temperatures of the two liquids are more easily and precisely discriminated if we immerse the fingers successively rather than simultaneously. The difference is less distinct if we use the fingers of both hands simultaneously; and it becomes very obscure if we use two adjacent fingers to explore both liquids at the same time.

My opinion is that finer discrimination is hindered by the merging of the two impressions. This explanation seems to me more probable than the one depending on the inability of the mind to attend to different things simultaneously. For if you explore both liquids at the same time but attend to the two fingers alternately rather than simultaneously, you still do not perceive the temperature difference as clearly as if you immerse the fingers successively.

This is very remarkable, because you might be inclined to suppose that discrimination would be even more impaired by the successive perception of impressions. After all, the first sensation must be held in the memory, and perhaps renewed by the strength of the imagination,

* The table below is identical to the one on p.[135] of *Der Tastsinn*.

so that it may be compared with the second sensation. It might be thought that it is easier to compare two simultaneous sensations with each other than to renew a previous sensation in the imagination and compare the renewed sensation with another new one. However, in both this example and other situations, we are better able to compare past with present sensations than two simultaneous sensations.

[130] I noticed the same thing when comparing the weights of two objects. Indeed, weights are very difficult to compare if they are placed simultaneously on one of the hands; but the comparison is easier if each weight is placed on a different hand, and easiest if they are placed successively on either one of the two hands.

The way in which the merging of two sensations hinders comparison is also clearly seen in the organs of taste and smell. Touch two places on the tongue simultaneously with two brushes, one of which has been moistened with dilute acid and the other with a sweet solution: if you attend to the two brushes alternately you will perceive different tastes, but not clearly, because the sweet and sour tastes merge in some way. You will distinguish both tastes much more clearly if you apply the acid and sweet brushes alternately to the tongue. Fill two small phials with different odorous substances, and bring them simultaneously to the nostrils. The sensation of the two odours will not be as clearly distinguished as when single phials are brought separately to the same nostrils.

A small phial containing a solution of hydrothionica was brought to a naive subject's right nostril, and another one filled with oil of turpentine was simultaneously brought to his left nostril. The sensations of the odour of both substances merged so thoroughly that he could not distinguish which nostril perceived which odour. However, you must take care not to use substances such as ammonia, which produce a strong irritation as well as an odour.

It has already been noted by Delezenne*[9] that two sounds produced at the same time are compared less easily and precisely than those that reach the ears in succession. Delezenne indeed demonstrated with very accurate experiments how acutely the human ear perceives an extremely small difference between two sounds. He says that musicians with a sensitive and well-trained ear can even perceive an interval between two tones as small as $\frac{1}{4}$ comma, $(81/80)^{\frac{1}{4}}$, i.e. a difference between two tones whose vibrations differ in the ratio of 645 to

* Charles Eduard Joseph Delezenne (1776-1866) was a physicist at the University of Lille, who worked on acoustical and electrical problems.

643*. [131] A subject with less musical experience can at least discriminate a difference as small as ½ comma, $(81/80)^{\frac{1}{2}}$, i.e. a difference between two tones whose vibrations differ in the ratio of 161 to 160. *However, pitch discrimination was as fine as this only for successive sounds. A much larger difference between the same sounds is not perceived if they arrive simultaneously.*

Vision is the only exception to this rule. We can compare the length of two parallel lines most accurately if they are close enough together to be seen simultaneously. Discrimination is worse if the lines are too far apart for simultaneous vision and the eyes must be turned from one to the other.

We shall now go through individual cases and demonstrate first, by repeated observations on many subjects, that *two weights are compared better if they are placed on the fingers successively rather than simultaneously.*

For this purpose I used two composite weights, each consisting of a combination of smaller weights. The weights took the form of vessels prepared so that the smaller ones might be contained by the larger. Within the first four-ounce weight was included a second two-ounce weight, and in this was a third one-ounce weight, and in this a fourth half-ounce weight, and so on. The under surface and overall shape of the same weights taken from the two sets was almost identical.

Two weights were warmed so that their temperature was almost the same as that of the fingers. They were then placed alternately and repeatedly on my fingers without my seeing them, so that they always touched the same place and were well supported by the fingers.

We first tried to discover the sensitivity and accuracy of weight discrimination if the weights were placed simultaneously on the fingers of the same hand, and the hand was supported and held still so that I could judge the weights only from their pressure, and not from the muscular effort needed to lift them. [132]

---

* Weber is referring to the comma of Didymus, or syntonic comma, which is equal to 0.22 of an equal-tempered semitone. It is the difference between a natural third and a Pythagorean third. The ratio of 81/80 is 1.0125, while the ratio between the frequencies of an equal-tempered semitone is 1.0595; the former is 0.22 of the latter. The ratio of 645/643 is 1.003, or the fourth root of 81/80; while 161/160, or 1.006, is the square root of 81/80.

EXPERIMENTAL SERIES 1

The difference threshold for two weights placed on the fingers of the same hand.

(a) Results when one weight was placed on the second and third finger, and another was *simultaneously* placed on the fourth and fifth finger of the same hand, when the hand was supported and immobile:

Weight placed on 2nd & 3rd finger, in half-ounces	Weight placed on 4th & 5th finger, in half-ounces	Judgement of which weight heavier
15	9	true
15	10	true
$11\frac{1}{2}$	15	true
$11\frac{1}{2}$	15	false

(b) Results when the first and second weights were placed *successively* on the abovementioned fingers, when the hand was supported and immobile:

15	12	true
13	15	true
15	14	true
$8\frac{1}{2}$	8	true
$12\frac{1}{2}$	12	true
$12\frac{1}{4}$	12	false
15	$15\frac{1}{2}$	false
15	$15\frac{3}{4}$	false
$14\frac{1}{4}$	15	false
$14\frac{1}{2}$	15	false
14	15	true

(c) Results when the weights were placed *simultaneously* on the abovementioned fingers, and held in a raised hand: [133]

11	15	true
15	12	true
15	13	true
14	15	false

(d) Results when one weight was placed on the second and third fingers, and then another weight *successively* on the fourth and fifth fingers, and the weights were held in a *raised* position:

14	15	certainly true
15	$14\frac{1}{2}$	false

(e) Results when the weights were placed repeatedly *in succession* on the second and third finger (not the fourth and fifth), and were held in a *raised* position:

15	14 ½	true

Several conclusions may be drawn from these experiments. First, weight discrimination is most sensitive and accurate if the objects are placed *successively on the same finger* and are *held with the hand in a raised position:* for under this condition a weight ratio of 29:30 is correctly discriminated (see experimental series 1.e.). Next, weight discrimination is also very accurate if the weights are placed *successively on different fingers of the same hand* (e.g. first on the second and third finger and then on the fourth and fifth), and are *held on a raised hand:* for in this condition a weight ratio of 28:30 is correctly discriminated (see experimental series 1.d.). Then, weight discrimination is poorer if the weights are *placed simultaneously* on those fingers and *held in a raised hand:* for weight discrimination is no finer than the ratio 26:30 (see experimental series 1.c.). Further, we are better able to discriminate between two weights if they are placed *successively* on our fingers rather than *held simultaneously on a raised hand:* for in the former case the ratio between the weights at the differential threshold was 28:30, or in another trial 24:25 (see experimental series 1.b.), but in the latter case it was 26:30. [134] Finally, discrimination is poorest if the weights are placed *simultaneously* on different *fingers of the same hand*, and are *not lifted* by the hand: for the difference was not perceptible unless the ratio of the weights was 30:20 (see experimental series 1.a.).

EXPERIMENTAL SERIES II

The difference threshold for two weights placed on the fingers of both hands.

Weight on 2nd & 3rd finger of right hand, in half-ounces	Weight on 2nd & 3rd finger of left hand, in half-ounces	Judgement of which weight heavier

(a) Weights placed *simultaneously* on the two hands, which were supported and immobile:

---

* The column heading must be understood to refer to the second weight rather than the fourth and fifth fingers.

11	15	true
7 ½	5 ½	false
5 ½	7 ½	true
12	15	false

(b) Weights placed *successively* on the two hands, which were supported and immobile:

15	14	true, but difficult
14	15	true and certain

(c) Weights placed *simultaneously* on the two hands, and then raised simultaneously:

14	15	true but uncertain
15	14	definitely uncertain
15	13½	true but difficult
13 ½	15	true and certain

(d) Weights placed *successively* on the left hand, and held in a *raised* position: [135]

15	14½*	true
14	15½	true

(e) Weights placed *successively* on the right hand, and held in a *raised* position:

15	14½*	true
14	15½	true
15	14½	true
15	14½	true
14	15½	true
14	15½	false
15	14½	true
15	14¾	false

From these experiments it should be concluded that a weight difference was noticed best if the objects were placed *successively* on the *same* hand and held in a raised position; for in this case the ratio of the weights could be 30:29 (see experimental series II.d. and e.). The difference was recognized less distinctly if the weights were placed *simultaneously on the two hands* and *held in a raised position;* for then the difference was not perceived unless the ratio of the weights was 30:27 (see experi-

* The column heading must be understood to refer to the second weight rather than a different hand.

mental series II.c.). It was perceived a little more accurately if the weights were placed on *the two hands successively*, but *were not lifted;* for then they were correctly estimated even if the ratio of the weights was 30:28. Finally, a difference was most difficult to discriminate if the weights were placed *simultaneously on the two hands*, and the hands were supported and immobile; for under this condition the difference was not noticed unless the ratio of the weights was at least 30:22 (see experimental series II.a.).

I also performed similar experiments on other subjects, e.g. on a fourteen year old youth.

OTHER EXPERIMENTS OF THE SAME TYPE [136]

On the difference threshold for two weights placed on the fingers of the two hands.

Weights placed on 2nd & 3rd finger of right hand, in half-ounces	Weights placed on 2nd & 3rd finger of left hand, in half-ounces	Judgement of which weight heavier
(a) Weights placed simultaneously on the two hands, which were supported:		
15	11	true
$5\frac{1}{2}$	$7\frac{1}{2}$	false
15	11	true
15	12	false
15	12	true
15	12	true
Similar experiments carried out on another occasion:		
15	$13\frac{1}{4}$	true
15	$13\frac{1}{4}$	true
15	$13\frac{3}{4}$	false
15	$13\frac{3}{4}$	true
$13\frac{3}{4}$	15	true
14	15	false
14	15	uncertain
(b) Weights placed successively on the two hands, which were supported and stationary:		
14	$15\frac{3}{4}$	always true
14	15	true

When the same experiment was repeated three times with the weights and hands interchanged, a correct judgement was always made as to which was heavier.

(c) Weights placed simultaneously on the two hands, and then lifted on the hands at the same time:

$15\frac{3}{4}$	14	uncertain
$15\frac{1}{4}$	14	uncertain
$15\frac{3}{4}$ [137]	$12\frac{3}{4}$	true
$12\frac{3}{4}$	$15\frac{3}{4}$	true

The same experiment was often repeated with the same results.

(d) Weights placed successively on the left hand, and held in a raised position:

14	$15\frac{3}{4}$*	always true
$15\frac{3}{4}$	15	true
$15\frac{3}{4}$	$15\frac{1}{4}$	4 out of 6 trials correct

In the experiments I have just reported, it was concluded beyond all doubt that a subject can compare the pressure of two weights much more easily, and judge their difference much more accurately, if they are perceived successively rather than simultaneously. For two impressions made *simultaneously* on the touch-sense mingle together and confuse the sense in some way, thus obstructing the clear and distinct perception of other things. The same thing also happens during temperature discrimination, and with the impressions made on the other senses, with the sole exception of vision.

I shall now enquire more precisely into the nature of the eye. Is vision really an exception to the rule that simultaneous impressions are more difficult to compare than successive ones? Eyes have the most acute vision in a particular structure: an object that falls on the axis of the retina is distinguished very clearly, but objects located beside it lose clarity the further from the retinal axis their image falls. If we observe very carefully, we notice that the part of the retina where the images are clearly distinguished is very small. So if we wish to perceive any line accurately and take in its length with the eye, we move the eye so that individual parts of the line fall on the retinal axis in turn. We get to know the length of the line by the movement of the eye, of which we are conscious, in the same manner as we discover the length of an object by movement of the finger. If we have to compare two lines, we estimate the first one and then turn the eye to the other one: the image of the second line is now placed on the retinal axis, and is seen as clearly as the first line. [138] It may be contended, then, that we never receive two impressions completely simultaneously, but in very quick succession; and the eye is formed so that it can perceive single parts of an object in turn. This is more true of vision than any other sense.

---

* The column heading must be understood to refer to the second weight rather than a different hand.

If we perceive two things in turn and compare them, we must compare a present with a past impression. It is therefore obvious that the past impression is recalled into memory, or is revived in us, along with the present impression; it is then observed so clearly in the mind that it is compared with the present impression even more accurately than are two present impressions that are attended to in succession.

However, this clearly seems inconsistent with the explanations commonly given by psychologists about the difference between present impressions of objects on our organs and past impressions renewed by the power of the imagination. For all psychologists agree that past impressions revived by the imagination are much less strong and clear than present impressions. For this reason people never take perceptions revived by the imagination to be the perceptions of things actually present - except for the insane, who confuse the two.

All this is no longer inconsistent if we assume that the human mind never compares two present impressions on the organs, but always a past impression, repeated in the mind, with a present one. With this assumption, it is clear why two weights that press both hands simultaneously for a long time are not compared as accurately as with successive presentation. Part of the reason for this is that the resuscitation may be somehow impaired if the pressure of the weights persists throughout the time in which we recall its sensation into memory. It is also partly because the merging of the two impressions makes it more difficult to observe the second weight, if both weights touch us at the same time.

[139] We come now to a very recondite topic, common to physiology and psychology - the doctrine concerning impressions that are made on our senses and later revived in the mind. My observations may help to clarify the topic, at least from one point of view.

We clearly do not know what happens inside us when a past impression made on the sense is revived. Physiologists and philosophers believe in a hypothesis that is certainly very probable, but unproven. This hypothesis is that things to be perceived excite a movement of the nerve fibres in the sense-organ, and the motion is propagated from there to the brain*, the prime instrument of the mind, where it is perceived by us. Furthermore, the same movement, though weaker, can be revived by the power of the imagination, either in the brain or in the brain and nerves.

I should like to state my own observations concerning the resuscitation of perceptions, after dismissing these and other hypotheses.

---

* The text has *crebrum*, which must be a misprint for *cerebrum* (correctly spelt in the remainder of the passage).

First, then, there must be no confusion between a revived perception of this type and an impression remaining in the senses for some time, or lasting after the cause of the impression is removed. For we are aware of the latter continuously, and do not distinguish it from an impression made by a present external object. Embers burning in the dark in a brisk circular motion arouse an image of a present circle of fire. This is because the rays of light fall on adjacent parts of the eye, and the impressions made on each point do not fade immediately. Perhaps the perception of a lasting effect of light is not distinguished from the perception of light itself. Rather, this perception is unitary and continuous.

If a weight currently on the hand is compared with a previous one, the finished sensation from the absent weight is renewed with the perception of the second weight: it is recalled to memory so clearly and distinctly that we perceive the weight difference accurately.

It should also be noted that in that comparison the sensation of weight is not brought to mind voluntarily, but through the perception of the second weight itself. [140] I often tried unsuccessfully to hold a previous weight sensation in the memory, or to revive it by the power of the imagination. However, during the time between the first and second impression I certainly received no sensation from the weight; yet as soon as the second weight was placed on the same hand or the other hand, I felt very clearly which weight was heavier.

Next, it must also be mentioned* that in this comparison no complete sensation of the previous impression is revived so that the second weight is compared with the first in our imagination; but, rather, the sensation of the second present weight obscures the revived perception of the first one, so that we perceive nothing of the first weight except its discrepancy from the second. It is the same as with two lengths** that almost cover each other except at the front; we see the whole of the length facing us, but only the part of the underneath length that is longer than the one in front. Finally, I am adding three propositions that have been proved by experiments.

The shorter the time interval between the perception of the first and second weight, the more accurately the perceptions are compared with each other. If the temporal interval is too long, the comparison becomes uncertain, or is definitely impeded.

---

* The text has *non praemittendum,*, which must be a misprint for *non praetermittendum*.

** *mensuris*, "measures".

If there is a larger difference between the two weights, then a larger temporal interval can also elapse between the two observations without impairing the accuracy of the comparison.

The perception of heavier weights is not impressed more firmly in the memory than that of lighter weights. That is, provided the geometric proportion between the weights remains the same.

I used two weights, one of 15 and the other of $14\frac{1}{2}$ half-ounces. I placed them successively on the second and third fingers of my colleague A., placing the second weight on the same spot as the first.
[141] My colleague possessed an acute sense of touch: he certainly perceived the difference in weight if the second weight was placed on the same fingers after one or two seconds, but not after 5 or 10 seconds.

Similar experiments were performed with weights of 14 and 15 half-ounces on my colleague B., and they showed almost the same thing. He perceived the difference clearly if one of the weights was placed on the same fingers 2, 5 or 10 seconds after the other, but not after 15 seconds.

It should also be noted that colleague B. did not possess such an acute sense as colleague A. A. was correct four times in perceiving a difference between $14\frac{3}{4}$ and 15 half-ounces, and never failed to recognize a difference between $14\frac{1}{2}$ and 15 half-ounces. B., however, could only distinguish between 14 and 15 half-ounces; while a third colleague C., in the same tests, could not even distinguish clearly between 15 and 14 half-ounces, but only between $15\frac{1}{2}$ and 14. If these weights were placed on the same fingers with a 10 second interval between the first and second weight, the difference between them was certainly not recognized; but with a 5 second interval the discrepancy was at least recognized, though not confidently. Similar experiments were also performed on other subjects, and on myself, with the same result. Slightly greater differences between the weights were correctly perceived, even with an interval of 30, 60 or 110 seconds between the first and second weight.

The same observation was made on myself, for example, when unidentified weights of 15 and 12 half-ounces were placed on the second and third fingers of the same hand, and were held by me in a raised position. In this case 60 or even 120 seconds elapsed without the weight judgements becoming incorrect.

If lighter weights in the same proportion were used, so that unidentified weights of 4 and 5 half-ounces, for example, were placed successively on those fingers, 60 or even 90 seconds could elapse without impairing the judgement. However, 120 seconds were enough to make the comparison uncertain.
[142] I was very surprised that I could, while fresh,

correctly perceive weights in the ratio of 15 to 14½ half-ounces, even if 15, 20, 25, 30 or even 35 seconds had elapsed between the two weights. However, it should be mentioned that in experiments of this type I repeated the same test three, four, five and even six times before I could judge the relation between the weights. With an interval of 40 seconds, my judgement between the same weights was uncertain and incorrect.

Many of the statements about tactile sensitivity for weight discrimination also apply to vision, when comparing the length of two lines. It is an easy experiment to examine visual sensitivity for size discrimination and compare it with tactile sensitivity for weight discrimination.

I cut up a sheet of writing paper of the most usual size into eight equal pieces*. I drew a straight and even line on each piece, taking care that all the lines were of equal thickness and blackness but of different length. The shortest line was 100 millimetres, the next 100½ mm., and the next 101 mm. In this way different lines were drawn up to a length of 105 mm.

Two cards were then placed beside each other, and shown to the subject whose visual discrimination I wished to examine. Subjects who work hard at the art of drawing, and so achieve very fine vision, were able to discriminate between perpendicular lines of 100 and 101 mm. When the experiment was repeated three to five times, they always judged the longer line correctly. However, even these subjects sometimes made mistakes when they were tired. Indeed, many subjects could not discriminate clearly between lines of 100 and 104 mm., but only between 100 and 105 mm. These experiments show that some subjects can discriminate quite clearly by vision a difference between two parallel lines of a hundredth part of the line, but others only a twentieth part.

We can compare vision and touch with each other, to discover whether tactile sensitivity for weight discrimination, or visual sensitivity for length discrimination, is the more precise. [143] The previous experiments make it clear that vision is more acute than touch. However, I was surprised that the difference between the two senses was not larger. For subjects who possess an acute touch-sense, but are not practised at weight discrimination, can distinguish clearly a thirtieth part of a weight; while those with excellent visual acuity can distinguish clearly a hundredth part of a line.

If the calculations and experiments of Delezenne

---

* *Dissecui chartam papyraceam scriptoriam magnitudine maxime consueta in octo partes aequales* ... The sizes of 'writing' sheets ranged from 61 x 49 cm. to 40 x 33 cm., with a middle size of 50 x 40 cm. When folded to give 8 leaves (octavo), the middle size was about 20 x 12.5 c.m. See P. Gaskell, *A new introduction to bibliography*, Oxford: Clarendon Press, 1972, pp.84-86.

(which I mentioned earlier) are true, hearing is more acute than vision or touch. For according to Delezenne, the division of a sound that has been recognized by the auditory sense of musicians is a 322nd part (the ratio being that of the number of oscillations constituting the sound); and by musically inexperienced subjects, a 161st part.

Many of the statements about the revival of previous weight perceptions by another weight sensation may also apply to the revival of a previous perception of a line by the sight of another line.

For example, we clearly recognize a fairly large difference between two lines, if the time interval between the two observations is fairly large. But if the difference is very small, the first impression of the line is obscured if the time interval is so large that the sight of another line cannot revive it clearly.

If the two lines for comparison were 100 and 110 mm. long, thus differing in the ratio of 10 to 11, and if the interval between the two observations was half a minute or even 70 seconds, discrimination was not impaired. However, it cannot be denied that with a 70 second interval the observation was so difficult that I was justified in believing that a longer interval would certainly have impeded it.

If the two lines for comparison were 100 and 105 mm. long, thus differing in the ratio of 20 to 21, an interval of 30 or 35 seconds between the two observations caused no obvious impairment in discrimination. [144] An interval of 40 seconds was long enough to obscure the impression made by the line seen first, so that the discrimination became uncertain.

Finally, if the two lines for comparison were 100 and $102\frac{1}{2}$ mm. long, thus differing in the ratio of 50 to $51\frac{1}{4}$, an interval of only 3 seconds rather than 5 or 10 could elapse between the two observations without impairing discrimination. If 4 seconds elapsed, the decision was more often correct than incorrect.

I found that in these experiments, too, the impression made by the first line did not persist after its removal, but ceased completely, so that it could not voluntarily be recalled to memory with accuracy, or reproduced in the imagination.

However, when the second line was inspected slightly later, the previous perception was revived, enabling very accurate length discrimination to occur.

When both lines were finally removed, both impressions were immediately extinguished so that it was impossible to compare their lengths correctly. For a present impression is much more easily compared with a past one than two past perceptions with each other.

Some further instructions should be added, so that the experiments that I have just described may be consistently and correctly repeated by others: the line was exposed to view for a specific time, such as 8 or 10 seconds; and I drew the lines on large enough pieces of paper (i.e. on octavo leaves)*, for a good reason. For there are two ways in which a pair of lines can be compared with each other: if they are far apart, as in the present case, we compare them by bringing into view first the whole of one line and then the whole of the other; but when the lines are parallel and close together we compare them in a different way. We first look at the lower end of each line, and by moving the eye in turn from the end of one line to the end of the other, we know by the eye movement the degree to which each line protrudes. Then, taking in both lines with the eyes simultaneously, [145] we go to the opposite ends: there we measure the protruding part of one of the two lines, in the same way as with the lower ends. We compare the lines much more accurately by this method than the other.

---

* *lineas laminis papyraceis tam magnis i.e. foliis octavis a me inscriptas esse.* See note on p. 105.

A SUMMARY OF OUR DISCOVERIES FROM THE EXPERIMENTS ON THE TOUCH-SENSE

There are two sorts of sensations aroused in the touch-organ.

Sensations aroused in the touch-organ are of two sorts. For we perceive (1) some pressure or traction, (2) warmth or cold. All these sensations seem to depend upon some movement aroused in the substance* of the tactile organ. However, the two motions are not the same. When an object presses on the skin, many skin particles are pressed simultaneously in almost the same direction; and if the skin is pulled, by grasping the hairs for example, the same particles are pulled *simultaneously* in almost the same direction. On the other hand, if a warm object gives heat to the skin or a cold object subtracts it, an expansion or contraction of *all* the individual particles must occur. This is because any particle *presses on neighbouring particles, or is pulled, in many directions*, and so *tension should arise between all the individual skin particles*, from which the special sensations of heat and cold certainly have their origin. For it is universally agreed that the addition of heat expands objects, while its subtraction contracts them.

Tactile acuity depends partly on the structure of the organ, and partly on movements of the organ made deliberately and consciously.

We often discover the shape of objects, and the distance between them, both by touch and by the movement of our organs which is obstructed by the objects. [146] This occurs, of course, because we are clearly conscious of the deliberate movement.

Experiments on tactile acuity were performed with the limbs sometimes immobile and sometimes freely moving, to avoid confusion between the type of tactile acuity that depends on the organization of the skin and the other type that arises from conscious and skilful limb movements.

---

* *materia.*

The separation and location of the touched parts of the body are distinguished accurately only in certain conditions.

We use various methods to gain quite accurate knowledge about the distance between different parts of the skin, and their location. We get to know this partly by vision, and partly by the deliberate and conscious movement of the hands when touching the parts. That is why people sensibly chose the thumb, the hand, the foot and the forearm* as units of measurement. For the complete length of a finger or a hand is very clearly impressed on the memory. It is unnecessary to compare this type of knowledge with vision; for the blind, who are very well practised, also possess very accurate knowledge of the distance between different parts of the body. They must have measured the distances only by such voluntary and conscious movements. We may enquire whether the knowledge of all parts of the skin is so firmly impressed on the memory that if we touch them without vision we can judge their location and separation correctly. Simple experiments tell us that we can do this only in the ends and boundaries of the parts, or else in those parts that are moved deliberately, frequently and in various ways. For example, if two objects touch the two edges of the hand, we can judge their separation correctly without vision. This supposes that the objects are separated by the width of the hand, and that we can recall the width of our hand to memory easily and clearly. The same thing happens if two fingers are touched. It is different on the mid-back of the hand: for example, if it is touched simultaneously by two legs of the compasses, separated by half a *thumb*, and placed longitudinally rather than transversely across the hand, we make an incorrect judgement about the distance separating the legs and the touched parts of the skin. [147] The points of the legs make only one impression and appear to touch the skin at only one spot; alternatively, if they make a double impression, the touched parts nevertheless seem to us so minimally separated as to be very close together.

Two objects touching the skin simultaneously seem to us to be separated by a smaller distance the lower the tactile acuity of the touched parts.

The finer the touch-sense of a given part of the body, the more accurate our non-visual judgement of the separation between objects touching the skin. This fact is even more certain when the objects are very close together. Surprisingly, the different parts of the body

---

* *ulna*, the forearm, elbow, or ell.

differ in this respect - the finger-tips, the back and palm of the hand, the forearm, the upper arm, the forehead, the cheeks, the lips, the back of the head, the chest, the back and other parts.

Sensitivity is inherently very low in the middle of the upper arm, the middle of the thigh, the back of the shoulder, and other places. When the compass points are separated by $1\frac{1}{2}$ Paris *thumbs* they produce a single impression, provided that they are arranged longitudinally along the upper arm and thigh. Under this condition the distance separating the two points is not clearly perceived. Points separated by 2, $2\frac{1}{2}$, or even 3 *thumbs*, and placed on those parts in the same way, seem to be separated by a very small distance. The better the tactile acuity of an area, the larger the space between the two points seems to be. On the finger-tips we perceive the separation of the points correctly even if they are separated by only one Paris *line*. When the compass points are so close that their impressions begin to merge into one, the location of the touch seems to have an oval shape; and when they are brought even closer together, the touched spot is similar to a point in which we cannot distinguish any longitudinal or transverse diameter. [148]

A remarkable phenomenon, which can be explained by the previous observation

A remarkable effect is obtained if someone else moves the compass legs continuously over the skin from places with poor sensitivity to others with good or very good sensitivity.

The compass legs should be separated by one *thumb* or $\frac{3}{4}$ of a *thumb*, and should be moved to the right cheek near the ear, with one point touching a higher and the other a lower place. The space between the two points will seem to you very small, or quite imperceptible, since tactile acuity in this place is very poor. However, the anterior part of the cheeks is much more sensitive, the lips even more so, and the middle part of the lips is the most sensitive of all: therefore the compass points will seem farther apart, the nearer they are moved to the middle of the lips. Even if the points remain the same distance apart, you will nevertheless seem to feel the points grow farther apart as they are moved over the skin. Actually, the points seem to move apart in curved rather than straight lines, since the sensitivity of the anterior parts grows by increasing rather than equal increments. The contrary is observed, if the compasses are moved across the skin from the mid-lip to the cheek near the other ear: the nearer the compasses to the ear, the closer together the points will seem, till eventually they appear completely, or almost completely, together. The sensation is also similar if you move the compasses so that the legs touch only the upper lip (see p. [59] & [60])*.

You will notice almost the same thing if the compass points are separated by 1/3 of a *thumb* and arranged transversely, and are moved quickly from the mid-forearm over the palm of the hand to the finger-tips: for although the space between them is always constant, you will nevertheless feel it growing larger the nearer the compass legs approach the extremities.

Hypotheses to explain why objects seem farther apart when they touch places with good acuity, but closer together when they are moved to less sensitive places. [149]

The following very probable hypothesis occurs to us, to explain the reason for the increase or decrease of the interval according to the differing degree of sensitivity: more nerve fibres terminate in more sensitive parts than in parts with a dull sense. When two impressions are made simultaneously on the same nerve fibre, only one common sensation arises; but when impressions are made on two fibres there is a double and differentiated sensation. The same nerve fibre undoubtedly brings sensitivity to many parts of the skin, and that is how such a small number of nerves is able to sensitise the whole surface of the skin; for you will not discover spots that definitely lack sensation - not even by pricking the skin with an extremely sharp point. Indeed, the spots where adjacent impressions are not merged are arranged in the skin so that in the very sensitive areas there are many of them, while in the parts with poor sensitivity there are very few. Such spots are easily distinguished: we are made aware of them by the prolonged use of the hands and other parts, and by varied contact. The more such spots are interposed on the skin between the two points of the compasses, the farther apart the points seem to be separated. Indeed the points make a single impression only if they touch one and the same place.

Different parts of the body differ in tactile acuity. The most sensitive parts are located in the face, the palm of the hands and the sole of the foot.

Experiments performed with the compasses offer us a method for measuring tactile acuity in different parts of the skin. We may note the distance between the compass points that is sufficient for distinguishing a double impression, or for discriminating between longitudinal or transverse orientations of the points in relation to the

---

* All page references within the text refer to the original page numbers of the Latin text, shown in square brackets in the translation.

limbs. [150] This shows us that the degree of tactile acuity that depends upon the structure is different in different parts of the body. For it will be possible to carry out a comparison of tactile acuity, not just in numbers representing the distance, but in *lines*.

The experiments that I described previously on p. [48]-[49] showed the following results: The touch-sense is especially acute on the volar surface of the hands, especially the finger-tips, and then on the tip of the tongue, then the lips, particularly at their edges, then the nose, especially on the edges of the nostrils and on the tip, then on the chin, next on the sole of the foot, especially on the tip of the big toe, on the heel and on the capitulum of the metacarpus of the big toe and of the smallest toe. Three areas of the human body, then contain parts with excellent tactile acuity: the face, the palm of the hand and the sole of the foot. Tactile acuity on the fingers is best on the last phalanx, then the middle phalanx, then the first. The upper arm is less sensitive than the *forearm*, the thigh than the shin, and the trunk than the head. The dorsal surface of the hands is much less sensitive than the volar surface, the lower surface of the tongue than the upper surface, the internal surface of the lips than the external, and the hairy surface of the head than the facial surface.

The middle part of the upper arm where the flesh of the biceps and triceps muscles is located, the middle part of the forearm, the middle part of the thigh and the same part of the shin where the soleus muscle is located, possess lower tactile acuity than the places next to the joints.

It is obvious from this that the ends and projecting parts of any place are more sensitive than the middle internal regions. However, the ear is an exception: even though it sticks out, it still has a poor tactile sense.

It is certainly beneficial that tactile sensitivity is most acute for those parts that protrude more, and which come into contact with objects most easily, and which we can move anywhere with precision.

Two objects simultaneously touching the longitudinal diameter of the arm or foot feel closer together than when they touch the transverse diameter of the same parts. [151]

It should be noted here that the compass points often make only a single impression if the line connecting both points is arranged along the length of the arm or foot, even though the impression is double and easily discriminable with a transverse orientation. This was observed, for example, if the points were 18 or 16 Paris *lines* apart on the middle of the upper arm, or 9 or 10 *lines* apart on the underside of the forearm, or if they were 12 or 16 *lines* apart on the middle of the shin. In these and other places the compass points seemed slightly closer together when applied transversely rather than longitudinally. (see p. [49]).

The same finding was observed in many other parts of the extremities, and also in the forehead and back of the head.

Hypotheses to explain why two objects touching the longitudinal diameter of the arms or feet feel closer together than when they follow the transverse diameter of the same parts.

In all these parts the nerve trunks run along the length of the body or of the extremities. It can therefore reasonably be supposed that the above fact depends upon *the direction of the nerves*, and the points make a fainter impression and seem closer together if the line joining both points *has the same direction as the nerves*. Certainly, this opinion is very strongly supported by the fact that the same thing is not always observed in the trunk, where many nerves (e.g. the intercostal nerves) have a transverse direction; in fact in several parts the contrary is observed. For in some parts of the trunk the compass points are better discriminated if they are applied longitudinally, but in other parts transversely. [152] However, I do not wish to imply that the cause of the difference lies in the larger parts of the nerves, but rather in the finer parts of the nerve endings.* Certainly, the endings have almost the same orientation as the larger fibres. Let us then assume that two impressions made on different parts of one nerve fibre excite only a single and not a dual sensation, but that one fibre communicates the ability to perceive to many points of the skin. When this is assumed, it is not absurd to suppose that the nerve fibres are perhaps arranged so that single fibres impart perceptual ability most strongly to most skin spots in the direction in which they proceed. However, in the direction at right angles to that of the nerves, the adjacent parts of the skin receive sensation from other nerve fibres, as the figure shown below makes clear.

c.

a. 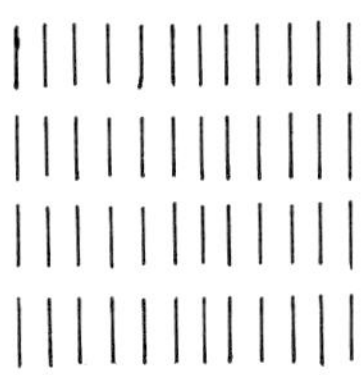 b.

d.

---

* *non... in nervis maioribus, sed in subtilioribus iam se finientium nervorum partibus ...*

Let a.b. be the transverse and c.d. the longitudinal direction of the arm or foot. Let the lines represent the endings of the nerve fibres, each of which provides sensation for many skin spots in the direction c.d. It is clear that in the same area many nerve fibres are arranged next to each other in the direction a.b., but few in the direction c.d. If this is conceded, it follows that when the compass legs are placed on the skin in the direction c.d. they generally make an impression on only one nerve fibre; but when applied in the other direction a.b., they usually prick different nerve fibres at the same time.

On the other hand, I do not deny that the variable effect of the compass points can also be explained by a slightly different cause than the direction of the nerves. [153] When the compass legs are applied in the same direction as the line of the nerves, they touch more easily those fibres that arise from the *same* branch of the nerve; but when applied in the other orientation they more easily touch nerve fibres originating from *different* trunks and branches. Stimulation of two nerve fibres originating from different nerves is more easily discriminated than stimulation of the fibres arising from the same branch.

However this may be, the following observation is quite certain: *when the compass points are placed on the arms and feet, whose nerves run along the length of the limbs, they are perceived more distinctly and seem farther apart if they are arranged transversely rather than longitudinally in relation to the limbs.* *(See p. [49]).*

Two objects touching the left and right side of the body simultaneously seem farther apart, and are more clearly distinguished, than objects touching only one of the sides.

There is another equally certain phenomenon, which depends on the origin and course of the nerves, and which I noticed in the mid-line of the anterior and posterior surface of the trunk and head, where the human skin divides into the right and left parts. *If the compasses are placed on the skin so that the two legs touch opposite sides of the body, the legs are more clearly distinguished and seem farther apart than if they both touch the same side.* This is a most remarkable fact. The nerve endings are of course located on the right and left of this line, and some of them originate from the right side of the brain and spinal medulla, and others from the left side. If, then, the phenomenon that was explained previously depends on the ordered arrangement of the nerve endings in the skin, the present phenomenon depends all the more on the place in the brain and spinal medulla where the beginnings of the nerves originate.

Two objects feel farther apart and are more clearly distinguished if they touch parts that we often deliberately move apart and together again, than if they press immobile parts. [154]

Here is another phenomenon that depends upon the conscious movement of the parts: *the compass legs feel farther apart and are more clearly distinguished if the two parts that they touch are ones that we often move deliberately to and fro, rather than immobile ones.* (*See p.* [*60*]).

We have often measured the distance between these parts when we draw them apart deliberately. This has given us a clearer knowledge of their location, and has also made us used to thinking the distance between them larger, even if they were not moved apart.

This is observed in the fingers, for example. If the compass legs touch only one finger, they do not feel as far apart as when they touch two adjacent fingers. The same thing happens in our gums: if the gums of the upper and lower jaw are pricked simultaneously by the compass legs, the impressions are clearer and seem farther apart than if the gum of one of the jaws is touched by both legs. I return now to a more precise description of tactile acuity in different parts of the body.

The whole of the head possesses better tactile acuity than the trunk and its individual parts. The parts are less sensitive the farther away they are from the extremities of the mouth, which have the best tactile acuity.

The *hair-covered region* is the most remote of these parts, but it still has a more acute sense than the neck. Nor indeed is the touch-sense of exactly the same grade over the whole of the hairy skin. *The crown* seems slightly less sensitive than the other parts of the hairy skin - the temples near the forehead, and the back of the head. *The tip of the tongue* and *the lips* are quite the opposite, being far more acute than the other parts of the head. In different parts of *the face*, the touch-sense loses acuity the farther away the parts are from the opening of the mouth and from the mid-line of the face. *The ears*, then, and *the part of the face next to them*, possess a rather dull sense. [155] In *the internal surface of the lips* the sensation is much less clear than in the *external* surface; and it becomes even less clear in the parts farther away from the edge of the lips. The touch-sense is therefore poorest near the gum. Pain is, of course, easily aroused in *the gum* itself; but we are not clearly aware of things touching it, and our perceptions there should be attributed partly to the impressions propagated to the teeth.

For in *the teeth*, when different ones are touched by the compass legs, we also have an obscure sensation (which undoubtedly resides in the pulp), such that if, for example, several teeth are touched simultaneously the interval is still distinguished.

The tip of the tongue is more sensitive than all other parts of the human body, including the finger-tips. The most acute sense lies only in the tip, and is confined to a very narrow area at the end.* This area, which is remarkable for its very acute sense, has a width of two or three *lines*. The cause of this very fine tactile acuity should not be sought in the papillae, which are of normal form and structure in this part of the tongue, but rather in the number of nerves, which are very plentiful in the tip of the tongue. The touch-sense deteriorates with the distance from the tip: this is true for *the upper surface of the tongue*, but even more so for *the lower surface*. However, the whole of the tongue is clearly one of the more sensitive organs of the touch-sense, and needs no further mention.

The pituitary membrane covering *the hard palate*, and the part of *the soft palate* that can be touched in the opening of the mouth, are more sensitive than the internal surface of the cheeks and the anterior surface of the gums. The chin also possesses good sensitivity, but under the jaw sensitivity deteriorates towards the neck (See p. [65]).

The trunk generally has a duller touch-sense than the head or limbs. Sensitivity is greater in the extremities of the trunk, in the region of the neck and anus, than in other parts of the trunk.

The touch-sense is less acute in the trunk than in the head and limbs. There is no part of the trunk that has such acute tactile sensitivity as the fingers, the palm of the hands, the tongue, the lips and the nose; indeed, there are only a few parts of the limbs that are less sensitive than any parts of the trunk.[156] The regions next to the head and anus have a finer sense than the other regions of the trunk. The tactile sense of the intermediate parts of the skin does not increase by equal steps as they approach these regions: for the uppermost part of the thorax seems to be slightly more sensitive than the lower part of the neck. The touch-sense near the umbilicus may also be slightly finer than on the pubic bone and under the xiphoid process. But these matters are not very certain.

It has already been shown above that the mid-line following the length of the trunk on its front and back

---

* *arctissimisque terminis circumscriptus est*. Cf. p. 43.

surface, where the nerves of the right and left sides come together, is remarkable for this reason: when the compass legs are placed there across the trunk, thus touching the right and left side of the body, they feel farther apart and are more clearly discriminated than when they touch only one side or are placed lengthwise along the mid-line itself.

I also noted that the touch-sense in the trunk was remarkable for another fact. When the compass legs are arranged horizontally on the trunk, in many places they are no better perceived and seem no farther apart than when arranged longitudinally; moreover, there are several places where the compass legs make an even clearer impression when arranged longitudinally rather than transversely. The orientation of the compass legs has a different effect in various parts of the trunk, because the cutaneous nerves proceed in some places along its longitudinal diameter and in others along its transverse diameter. It might easily be believed that the breasts, and especially their tips, have the finest touch-sense of all: for in that area there are numerous papillae, which physiologists have up till now considered as belonging to the organs with the more acute touch-sense. However, experiments on this matter that I conducted on males show the opposite, since the touch-sense is very poor there. The cutaneous papillae in this place therefore have some other use than sharpening the touch-sense. Finally, it should be mentioned that the anterior surface of the trunk is not superior in tactile acuity to the dorsal surface. (See p. [67]).

The touch-sense of individual places in the limbs is generally finer the nearer they are to the extreme ends. However, the touch-sense is more acute towards the joints, i.e. the connecting tissue (commissures) of the parts, than in the middle parts between the joints: for some of the latter have no better tactile acuity than many parts of the trunk. [157]

Among the parts of the body with the poorest touch-sense are the places with the largest supply of muscles, approximately in the middle of the upper arm, thigh, forearm and shin. This place is not precisely in the middle of the shin and forearm, but is nearer the joint of the knee and elbow. The articulated extremitics are more acute than the middle part of the upper arm, lower arm, metacarpus, thigh, shin and metatarsus. In many joints, the convex surface of the flexed joint seems slightly more acute than the concave surface. Superior tactile acuity occurs for the region of the patella in the knee, the region of the olecranon and condyles of the humerus in the flexed elbow, and for the acromion in the axilla. The upper arm is less acute than the forearm, the forearm than the hand, the thigh than the shin, and the shin than the foot, while the foot is less acute than the hand by a surprising degree. In the hand sensitivity becomes more and more acute towards the finger-tips. The internal surface of the upper arm and forearm does not differ from the external surface, nor does the posterior surface of the thigh and shin differ from the anterior surface, as regards tactile acuity for a constant procedure. However, the touch-sense is much poorer in the dorsal surface of the hand and foot than in the palm and sole. The grooves arising from the folding of the skin in the palm of the flexed hand, and those of the individual fingers, are less sensitive than the adjacent parts of the skin. (See p. [63]).

The direction in which objects press the skin is not perceived by touch.

There is no tactile perception of whether a stylus is in a perpendicular or oblique orientation when it is placed on the skin. This subject is worth attention, especially because there is as yet no satisfactory explanation as to how we perceive the direction of a ray of light by vision; and it is very important for us to investigate the same aspect of this question in experiments on the touch-organ. In fact, we generally assume that the stylus is advanced at right angles to the skin; and we judge the direction of objects touching us by a decision-process consistent with this assumption. (See p. [78]). [158]

We recognize with great accuracy the direction in which hairs are pulled; we do so not only by the sense of touch, but also by the sensations attending the movement of the muscles when resisting the pull of the hairs.

It must be mentioned that we make very correct and accurate non-visual judgements about the direction in which the hairs are pulled. If someone grasps a small bundle of my hairs and pulls them, I feel the direction of pull with surprising clarity, and can indicate it to others with my finger.

Actually, I have discovered by experiments that the cause lies both in the sense belonging to the hair bulbs, and in the muscles, especially in the subcutaneous ones. The skin is of course moved by the pull of the hairs, but the subcutaneous muscles resist this movement. Since we are conscious of the direction of this muscular movement, we recognize the direction in which the hairs are pulled. Because this knowledge depends both on the skin sense and on the consciously perceived movement of the cutaneous muscles, it becomes uncertain when the skin movement is completely impeded. Place the head on a couch and make it immobile; then press the skin to the skull in the area where the hair is pulled, and so immobilize this part too. When this is done, there is no longer any reason why the subcutaneous muscles should resist the pull of the hairs, because the skin is no longer moved by the pull of the hairs. When this muscular movement ceases, we also lose most of the ability to recognize the direction along which the hairs are pulled.

We understand from this a new use of the subcutaneous muscles; and we discover the reason why the sense through which we perceive the direction of hair pull should be poorer in the parts of the skin that lack subcutaneous muscle. Animals certainly feel the direction of the winds moving the hairs with great precision, by means of the subcutaneous muscles spread over most of the body. However, their thicker and longer hairs are often located around the mouth and are more useful than the others as instruments of the touch-sense: they are arranged so that they can be moved *individually*. [159] For this reason their roots are contained in very big capsules constructed from muscular fibres, which the other hairs lack. The individual hairs, then, should be compared with probes, which enable the animals to feel not only the contact and the pulling, but also the direction of the pressure and of the pulling.

The muscular fibres in the hair capsules of the seal were described by RUDOLPHI*(RUDOLPHI *Grundriss der Physiologie* B.II, p.82) and in rabbits by RAPP** (W. RAPP, *die Verrichtungen des fünften Nervenpaares, mit 3 Tafeln. Leipzig* 1832. 4. p.20). According to Rapp, these hair muscles seem to accept branches from the facial nerve, but the hair bulbs are constructed with branches from the infraorbital branch of the trigeminal nerve. For when the trunk of the facial nerve in rabbits was mechanically irritated, those hairs trembled together with the facial muscles; but when the infraorbital nerve was irritated, no movement was observed.

The relative parts played by the cutaneous and muscular senses in perceiving the weight of objects.

The weight of objects is perceived in two ways: first by the touch-sense which resides in the skin; and then by the sense of the voluntary muscles, by which we recognize the effect involved in overcoming the resistance of weights and other objects, and lifting them up.

---

* Carl Asmund Rudolphi (1771-1832) was a Swedish scientist who became professor of anatomy and physiology at the University of Berlin from 1810.

** Wilhelm von Rapp (1794-1868) was a German physician.

The two methods are almost always used together. To reveal the relative contributions of the touch and muscle senses to weight perception, experiments must be carried out in which the weight of the objects placed on our organs is first judged solely by touch, excluding all use of the muscles. For this purpose, let two weights be placed on a supported part of our body, and let us compare them with each other. For under this condition the weights are perceived only by touch. As one of the weights is progressively increased or decreased, there will arise a difference sufficiently great that they are always discriminated correctly by touch, even when the weights change place. [160]. The smaller the difference that can be recognized by touch, the more acute the tactile sense.

Once it is known what weight difference is necessary for tactile discrimination with the limbs immobile, then both methods of judging weights (both touch and the muscular sense) must be used together.

The same experiment must be repeated, then, in such a way that the weights are lifted simultaneously by the limbs. When this is done you will notice that the discriminable differences are certainly smaller.

It is possible to increase the lighter weight by a certain amount whilst preserving the discrimination. This amount will show the extent to which weight discrimination is more accurate when the weights are raised by the limbs rather than placed on the supported and immobile limbs. My experiments on this matter show that most subjects have much finer discrimination when they lift the weights, though not all subjects show the same degree of improvement. In an experiment (see p. [132] a.c.), when the hands were supported and immobile, 15 and 10 oz. were clearly discriminated from each other, but not 15 and 11¼; but if the hands were lifted, 3 oz. could be added to the lighter weight whilst preserving the discrimination, so that 15 and 13 oz. were distinguished. In the first case, then, the just discriminable difference was 5/15, while in the second case it was only 2/15. In another experiment (p. [137] a.c. when the hands were supported and immobile, 15 and 11 oz. were distinguished, but not 15 and 12; but with lifted hands 2 oz. could be added to the lighter weight whist preserving the discrimination, so that 15 and 13 oz. were distinguished. In the first case, then, the just discriminable difference was 4/15, while in the second case it was 2/15. In another experiment (see p. [137] b.d.) 15 and 14 oz. could be distinguished, but not 15 and 14½, with the hands immobile and supported; but when the weights were lifted, half an ounce could be added to the lighter weight whilst preserving the discrimination, so that 15 and 14½ oz. could be discriminated. In the first case, then, the just discriminable difference was 1/15, while in the second it was 1/30 (see p. [86] & [132]). Less practised subjects (see experiments described on p. [90]) could generally discriminate between weights of 32 and 29 ounces if they were placed simultaneously on both hands when supported and immobile; but when the weights were lifted by the hands, the difference between 32 and 29 oz. could generally be recognized. [161] The just discriminable difference, then, was 11/32 in the first case but 3/32 in the other case. The reason for the poor discrimination in these experiments when the weights were not lifted lies in the fact that the same weight affects the two hands differently: it generally causes a stronger pressure sensation in the left hand than the right. Weight discrimination by this method is therefore disturbed to some extent. But when the weights are lifted by both hands, the sensations arising from the contraction of the muscles on the right and left side do not seem very different; and so the comparison of weights by this method is more accurate.

Weights arouse a stronger pressure sensation in those parts of the skin where our tactile acuity for the separation of the compass legs is best.

Equal weights do not cause the same pressure sensation everywhere when placed on supported and immobile hands or arms: they arouse a stronger sensation in the parts with many nerves, which therefore possess a finer sensitivity; while they arouse a weaker sensation in the parts with fewer nerves, which therefore possess poorer sensitivity. The same thing would happen if a man lay on his back with one weight on his mouth and another on his forehead; or if one weight were placed on the palm and the other on the back of the hands. The same thing is found in all parts of the body which differ in sensitivity, if each part is supported and immobile. However, if you place a lighter weight on the mouth, or a heavier one on the forehead, at a certain difference you will excite an equal sense of pressure*. The reason is that the impressions transferred to the brain through the nerves are more forceful both if the force making the impression is greater, and if the impression is propagated to the brain through more nerves. Therefore a less forceful impression propagated to the brain through many nerves excites a sensation as forceful as a more forceful impression transferred to the brain** by a few nerves. [162] It is therefore clear that weights to be compared with each other must always be placed on equally-sensitive organs; if you ignore this rule, your more precise experiments will contain surprisingly large errors. For example, the last phalanges of the fingers are more sensitive than the middle ones, and the latter are of course exceeded in sensitivity by the former; therefore weights for comparison must be placed on the same phalanges, for otherwise weights nearer the finger-tips feel heavier. I tried in

* The text reads *aequalem sensus pressionis excitabis*, but *sensus* must be a misprint for *sensum*.

** The text has *crebrum*, which must be a misprint for *cerebrum*.

many experiments to discover and measure the degree of tactile sensitivity in many parts of the body, using the method in which I increased the weight on the less sensitive organ until the two weights produced an equal pressure. I was very interested to know whether the variations in tactile sensitivity in different places would be the same when measured by adding weights as when tested by the compasses. I generally found the same result with the two methods for examining tactile sensitivity. For the same parts that showed the best tactile sensitivity when examined by the compasses were also best when the weights were imposed. However, the difference in sensitivity did not appear equally large in both experiments, and the first method is preferable to this one (see p. [97]).

The sensation is similar, though not quite equal, if the weight lies on a larger rather than a smaller skin surface.

The sensation is similar but not quite the same if equal weights are placed on our organs, the weight with the larger base lying on the larger skin surface, and the one with the smaller base on the smaller skin surface. The more particles of skin a weight rests on, the less any particle is pressed, and *vice versa*. However, sensations aroused simultaneously in different but adjacent particles merge into one sensation: therefore the sensation grows stronger both when any particle is pressed by a stronger force, and when many adjacent particles are pressed simultaneously. That is why the sensations arising from two such weights become almost the same if the weights are placed simultaneously on the forehead for a fairly long time. [163] In the palm of the hands we immediately perceive the basal circumference of the weights more clearly than in the forehead; the sensation that we know in objects noted for their high specific weight is aroused there more clearly from the weight with the larger base. (See p. [85])*.

A difference is perceived no more easily between heavier than between lighter weights.

If two weights are placed on the hands, and one weighs 32 oz., and the other is gradually reduced until we feel the difference between the weights by touch, we

* In that passage it is clearly stated that the weight with the smaller basal area feels like a denser metal. That is a correct example of the size-weight illusion (see Ross, H.E. When is a weight not illusory? *Q.J. exp.Psychol.* 1969 *21*, 346-355.). However, in the present passage Weber seems to have it the wrong way round.

learn the minimum ratio between the weights that can be recognized by touch. Now if we repeat the same experiment using weights of 32 drachms instead of the previous weights, i.e. eight times less, we observe that the minimum proportion still recognizable by touch is almost the same as in the previous experiment: for example, if weights of 32 and 26 oz. could be discriminated before, those of 32 and 26 drachms will now generally be discriminable. (See p. [91]).

We feel the temperature of objects most clearly in those parts where our tactile sensitivity for the separation of the compass legs and for the difference between weights is best.

We perceive a very small temperature difference between two objects more distinctly with the fingers than with other parts, and with the palm than with the back of the hand. The lips and the nose, though inferior to the fingers, also have a very distinct temperature sense. However, it is hard to compare individual parts of the body accurately with each other in this ability. Some parts, you see, are covered by a thicker epidermis or cuticle than others. The epidermis is not only totally lacking in sensation, but is also a poor conductor of heat. The thicker the epidermis, the greater the difficulty of penetration by heat: therefore parts possessing poorer tactile sensitivity but also a thinner cuticle are often affected by heat more quickly and strongly than other parts that have more acute sensitivity in other respects. [164]
So, for example, the cheeks have poorer tactile sensitivity than the lips and nose, but are more strongly affected by heat, for they are covered by a very thin cuticle. The palm of the hand is more sensitive, but the back has the thinner cuticle: if, then, you place both hands together in a wooden flask filled with cold water (+ $9\frac{1}{2}^{\circ}$R of heat), with the ulnar edge downwards and the radial upwards and the two hands slightly apart, the water touching the back of the hands will always seem colder at first than the water in between the two palms; but when 10, 15 or 24 seconds have elapsed, the sense of cold in the back of the hands gradually diminishes, while it increases in the palm so that eventually the water between the two hands seems colder than that touching their backs. The same holds for the sense of heat, if the hands are immersed in water of 38°R in temperature. Since the back of the hand is covered by a thinner cuticle than the palm, heat and cold affect the nerves of the back more quickly than the nerves of the palm, and so the sense of heat or cold arising from the hot or cold water is stronger at first in the back. But because the palm possesses finer sensitivity owing to its greater number of nerves, the heat and cold gradually overcome the obstacle of the thicker

cuticle and have a stronger and more consistent effect in the palm. (See p. [94], [123] & [125]).

Several physiologists have been mistaken, then, in holding that the sense of temperature belonged to a unique sense different from the touch-sense. For if the parts of the skin that are oustanding in the touch-sense also perceive temperature very acutely, it is likely that the same nerves perceive pressure and heat.

A smaller degree of heat received from a larger skin surface causes the same sensation as a larger degree of heat stimulating a smaller skin surface.

If someone puts his finger in warm water of 32 or 33 degrees, and at the same time puts his whole hand in warm water of 29½ degrees, the latter water feels warmer than the former. This effect depends on the fact that impressions from individual particles of water made in different spots of the skin merge into one impression. [165] The impression made on the nerves by heat is of course propagated to the brain, and there arouses a sensation that increases partly with the degree of heat stimulating the nerves, and partly with the number of nerves receiving it and propagating it to the brain. Experiments for the comparison of tactile sensitivity in different parts must therefore be carried out so that the touched parts are stimulated by the same degree of heat and are equal in surface area. (See p. [113]).

When a larger skin surface is touched by warm objects, we recognize a difference in their temperature more clearly than when a smaller surface comes into contact with the same object.

This effect arises from the same cause - the merging of individual impressions made on individual nerves. With one finger immersed, I could not feel the difference in temperature between two liquids of $20\frac{1}{4}$ and $20\frac{2}{3}°$ R, but with the whole hand immersed I certainly sensed it. A similar effect is also observed with the other senses. A minute difference between colours that is not distinguished on a very small surface will be perceived if the coloured surfaces are large enough. Similarly, taste and smell are clearer if the surface of the organs coming into contact with the test objects is sufficiently large. The sense of smell in many animals seems much more acute than in man, because the membrane containing the olfactory nerves is much larger. In dogs, for example, and other animals, the whole nasal cavity covered by the olfactory nerves is larger, and it is also divided into smaller spaces by many more bony scales covered by mucous membrane. This causes many more odiferous particles to come into contact with the membrane.

We use the touch-sense to perceive the quantity of heat that is either subtracted from or added to the skin in a certain time. [166]

Objects such as metals from which a large quantity of heat is quickly drawn by colder objects feel warmer than other objects of the same temperature, such as wood, which do not lose heat as quickly. The same is also true of the sense of cold. For the same reason, the same object often feels cold at the times when we are hot, and hot at the times when we are cold. For under the former condition heat is drawn from the skin, and under the latter it is added to it. (See p. [113]).

A very small difference in temperature between two objects is perceived very accurately if the temperature is moderate.

You should not believe that a very small difference in temperature between two objects is more easily perceived if the temperature is only just below the level at which a burning sensation is provoked. In fact, if the heat is sufficiently intense to cause pain, or nearly so, the touch-sense immediately becomes less acute. The ears also pick up very small differences between sounds more easily if the sounds are neither too high, nor too low, nor too loud. Similarly, as has been found experimentally, a moderate temperature that does not hurt the fingers is most accurately perceived. The sense is also true for that object temperature that is little different from the temperature of the hands.

The weights of two objects are more easily compared and more accurately discriminated if they touch the skin successively rather than simultaneously; and the same holds true for the different temperatures of test objects, which are more clearly distinguished when they stimulate the skin successively. The other senses are also subject to the same rule.

The reason for the obstacle to accurate comparison when two simultaneous impressions merge does not lie in the fact that the mind cannot deal simultaneously with different things. For even if you turn your attention alternately to different parts while examining the weight or temperature of two objects, you still do not perceive the difference as clearly as if you apply the fingers successively.[167] This fact is most certain, although you might easily have believed that two present impressions are compared and differentiated better than is a present with a past impression renewed by the imagination. The merging of two simultaneous impressions is also evident in other senses, especially in smell and taste. If you touch two parts of the tongue simultaneously with two brushes, one of which has been soaked in dilute acid and the other in a saccharin solution, you will actually perceive a different taste, though not clearly, if you turn your attention to each brush alternately. However, you will discriminate both tastes more clearly if you move the acid and sweet brushes alternately to the tongue. If a small phial full of oil of terebinth is moved to the left nasal aperature, and simultaneously another one full of a solution of hydrothionicus to the right, the two sensations merge so completely that you would not discriminate clearly in which side of the nose you smell which of the odours. According to Delezenne, sounds are also compared with each other much more accurately if they strike the ears successively rather than simultaneously. At first glance vision appears to be the only sense excepted from this rule; for we are best at comparing the length of two parallel lines if they are close enough to be seen together - better than when the lines are so far apart that they cannot be seen simultaneously, but the eyes have to turn from one to the other.

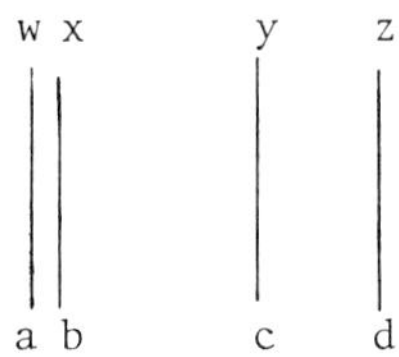

Lines a and b have the same relative lengths as lines c and d. Nevertheless, it is very easy to see that line b is certainly longer than line a*, but it is not so easy to be sure that line c is longer than line d; and the decision would be even less certain if you first inspected line c and then line d drawn on another page. [168] But you would be mistaken if you concluded from this that vision is an exception to my rule. The reason for the clearer difference in length between the lines a and b is not that their images always fall simultaneously on the most acute central part of the nervous tunic, while the centre of the eye must be turned successively to the lines c and d; rather, it is because we draw lines in our imagination through the ends of the lines w, x, y, z, and we conclude from the angles made with a horizontal line that the lines a and c are longer than the lines b and d. The farther apart the lines are, the smaller is the angle made with a horizontal line, and so a very small difference in the length of the lines strikes the eye with greater difficulty. The truth of this is also apparent

* According to the diagram, line a is longer than line b. The text seems to be in error.

from the fact that two very thin threads, or two hairs, are compared for thickness better if they are far enough apart than if they are as close together as possible. The same thing occurs when discriminating the colour of two very similar silk threads. The eye is formed in such a way that the mind is not as greatly perturbed by many simultaneous impressions as it is from the other sense-organs. Only those objects whose images fall on the centre of the retina are clearly distinguished; and this central area is so small that, with the eye stationary, objects are clearly distinguished only if their rays of light make a very acute angle with the axis of the eye. Therefore, when many things enter the eye simultaneously they do not disturb observation much by merging together. For the impression made in the centre of the retina is normally so strong that other impressions in other parts of the nervous tunic are obscured, especially if our attention is turned to the centre of the retina. However, if the impression made in the centre of the retina is very weak, the light rays from other objects that enter the eyes at the same time must be excluded from the eyes as far as possible. For this purpose we usually display very small objects against a black background*, so that we may inspect them carefully. Many things that we seem to comprehend simultaneously with our eyes we see successively, turning the eyes very quickly from one to the other. (See p. [129] and following).

Even the eye, therefore, is no exception to the common rule that things are more accurately compared with each other if they strike the senses successively rather than simultaneously. [169] In fact, the eye is organised so that we inspect everything successively. For there is only one part of the retina that is certainly suited for clear vision, and that part is very small and central: the images falling on the other parts of the retina make impressions that are less clear and strong. We therefore turn the eye so that the images of the things we wish to inspect fall successively on this central place. In this way we are less disturbed by the images of other things that are simultaneously in view.

If the palm of the hand has a temperature of 29 or 29½°R, water with a temperature of 29°R excites a sense of heat in the immersed hand, whilst with a temperature of 28°R it excites a sense of cold.

It might be supposed that the sensation of cold would always arise if an object draws heat from the skin, and the sensation of heat if heat is communicated to the

* *res exiguas ... nigris corporibus infigere solemus.*

skin; objects warmer than the skin, which communicate heat to it, would feel warm, while those that are colder than the skin and subtract heat from it would feel cold. However, experiments show us that water feels warm when it has a temperature of 29°R and touches skin with a temperature of 29½°R, and so is colder than the skin. The reason is certainly that less heat is subtracted from the skin in a given time by water of 29° than arises in the skin by metabolism*. Therefore heat must accumulate in the skin and a sensation of increased heat must be aroused. But water of 28°R seems to subtract so much heat from the hand that the warmth secreted in it is insufficient to preserve the same temperature. The temperature of the hand therefore decreases, and we are then affected by a sensation of cold.

The sensations of heat and cold differ from each other in a different way than do sensations of light and dark.

The sensation of dark is the sense of the absence of sufficient light for discrimination. Since light is never completely absent, the sensations of light and darkness differ only in degree. It follows, then, that one sense gradually changes into the other as the light increases or decreases: there is no middle stage, in which we are affected neither by a sensation of light nor by one of darkness. [170] On the other hand, such a middle degree certainly exists for the temperature of objects that arouse in us neither cold nor heat; but this range is confined by very narrow boundaries. The reason is that heat is communicated to our body from warmer objects and is subtracted from it by colder objects; but light is never subtracted from the eyes, only given to them. Cold and heat, then, exist as positive and negative numbers, the middle of which is a point of indifference; but light and darkness exist as greater or smaller positive numbers.

The most perfect darkness totally without light, if it were ever found, would be placed equal to zero. As for the rest, it is clear from the above discussion that the same objects with the same temperature can arouse in us a sense of cold or heat at different times. For since our skin varies in temperature, heat is sometimes communicated to it and sometimes subtracted from it by the same objects at the same temperature; and so a sensation sometimes of heat and sometimes of cold is aroused.

The sensation of burning arises from too strong an irritation of the brain.

An irritation of the brain caused by heat can become

* *quantum nutritione in cute nascitur.*

too strong and cause pain for two reasons: either because the degree of heat affecting the body is too strong, so that the material of the organs is sometimes destroyed, and is generally changed chemically; or because the area of the body simultaneously affected by heat is too large. The impression in the brain is of course stronger the more nerves carry the same impression to it simultaneously. This would happen if the major part of the skin were simultaneously affected by heat. Single impressions, then, which are carried together to the brain by the nerves, merge into one impression, which is stronger the more impressions merge together. Water with a temperature of 41°R does not excite a sensation of burning in one finger, even if it is immersed for a long time; but it affects the whole hand with this sensation if it is immersed for a long enough time. For the same reason, when parts possessing many nerves, such as the tip of the tongue and the finger-tips, are exposed to heat along with other parts, they are more strongly affected and feel pain more easily than the other parts.

Two simultaneous impressions, whether visual or tactual, merge as if into one if the parts of the visual or tactual organ that they strike are separated by an appropriate distance. [171]

If two parts of the skin touched by two objects are separated by not more than ¼ of a Paris *line*, we seem to be touched by only one object, even in the parts with the most acute tactile sense. If two parts of the mid retina are touched simultaneously by two rays of light, the same thing is observed; a double impression produces only one sensation. However, the interval between both places must be much smaller for the sensations to coalesce into one - actually less than 1/600 of a Paris *line*. The reason why simultaneous impressions merge together so easily in the skin and are taken as one is certainly because we differentiate only between those impressions that arise in different nerve fibres. Nerve fibres in the skin are much rarer than in the retina, which consists almost entirely of nervous medulla: it therefore follows that two simultaneous impressions made on neighbouring parts of the skin merge much more easily than in the retina. The much greater acuity of the eye in this respect may be indicated and expressed numerically as follows: if the acuity of the touch-sense is considered equal to the distance between the impressions to be discriminated, which is here equal to ¼, the acuity of the sense in the centre of the retina would be equal to 600; or what is exactly the same thing, the acuity of the touch-sense compared to the acuity of the central part of the retina would be in the ratio of 1:2400.

Concerning the smallest differences between impressions that can be recognized by the eye, the ear and the touch-sense.

Subjects who are well practised at weight comparison can distinguish a difference between two weights, when lifting them successively with the same hand, of 30 ounces or drachms from 29 ounces or drachms. Those who are slightly less practised or skilful can discriminate 14 from 15 ounces of drachms. When the hands are deliberately moved the touch-sense is so acute that a weight *difference* can be perceived that is equal to a *fifteenth or thirtieth part of the weights*.

[172] Subjects who are well practised at comparing the length of two lines could distinguish between lines of 100 and 101 mm. when drawn on paper and separated from each other by four *thumbs;* while less practised subjects could perceive only a much greater difference in length, such as between lines of 100 and 104 or 105 mm. The former subjects, then, could discriminate some difference in a *hundredth part of the length*, while the latter could discriminate *a twentieth part*.

According to experiments carried out by my brother Wilhelm, well practised subjects can compare two sounds differing in pitch (i.e. the number of vibrations) so accurately that they always perceive a difference if the number of vibrations is in the ratio of 200:201. They can perceive, then, a *two hundredth* part of the vibrations. According to Delezenne, the most expert musicians can even perceive a difference in sounds arising from vibrations in the ratio 322:321.

Vision exceeds all other senses in the clarity and subtlety of the impressions that it perceives, but it is certainly noteworthy that it does not exceed hearing in observing the differences that I mentioned.

Now to describe my pronouncements in a few words:

> we do not find weight discrimination *by touch* unless the difference between the weights is at least a fifteenth or thirtieth part;
>
> we do not find discrimination between lines *by vision* unless the difference between them is a hundredth part;
>
> lastly, we do not find discrimination between sounds *by hearing* unless there is a pitch difference of a three hundred and twenty second part of the vibrations.

*When noting a difference between things that have been compared, we do not perceive the difference between the things, but the ratio of the difference to their magnitude**

If two weights of 30 and 29 half-ounces are compared by touch, the difference is perceived no more easily than for weights of 30 and 29 drachms. [173] The difference between the weights is one half-ounce in the former case and one drachm in the latter. However, since a half-ounce equals four drachms, the weight difference is four times smaller in the latter case than the former. But since the difference is perceived no more easily in the former case than the latter, it is shown that what is perceived is not the absolute weight difference but the ratio. The same thing is also evident if we compare weights whose difference is so small that it is certainly not perceived. If two weights of 33 and 34 half-ounces are compared by touch, we do not perceive the difference; but that is not because the half-ounce difference is too small a weight to be perceived by touch (for we clearly perceive weights a hundred times smaller by touch), but because the ratio of the difference between the two weights is only a 34th part of the heavier weight. For it has been shown experimentally that skilful and practised subjects can perceive a difference if it is not less than a 30th part of the heavier weight, and the same subjects perceive the difference just as easily if half-ounces are replaced by drachms.

My statements concerning weights compared by touch also hold true for the comparison of lines by vision. For whether you compare longer or shorter lines, you will find that most subjects cannot perceive the difference if one line is smaller by a hundredth part. The difference in length between two lines is greater in absolute terms if you compare long lines, and less if you compare shorter lines. For example, if you compare a line 100 mm. long with one of 101 mm., the difference will be equal to 1 mm. If you compare a line of 50 mm. with one $50\frac{1}{2}$ mm., the difference will be equal to $\frac{1}{2}$ mm. The disparity is recognized as easily in the latter case, even though it is twice as small, because in both cases the difference between the two lines is equal to 1/100 of the longer line.

---

* The following passage is famous as the first clear statement of Weber's Law. Alternative translations appear in Titchener, E.B. *Experimental Psychology* II, part II, p.XVI, New York: Macmillan, 1905, (Reprinted in Rand, B. *The classical psychologists*, London: Constable, 1912; and in Dennis, W. *Readings in the history of psychology*, New York: Appleton-Century-Crofts, 1948); and in Herrnstein, R.J. and Boring, E.G. *A source book in the history of psychology*, Cambridge, Mass.: Harvard University Press, 1965. Translations of short passages from *Der Tastsinn* will also be found in these books.

I have not carried out experiments concerning sounds compared by hearing. However, from the experiments of Delezenne that I described, it is shown that very well practised musicians can just perceive an interval between two sounds as small as $\frac{1}{4}$ comma, or $(81/80)^{\frac{1}{4}}$, i.e. a difference between two sounds whose vibrations differ from each other as 645 to 643, or, which is almost the same thing, as 322 to 321. [174] These musicians, then, discriminated by ear 1/322 of the vibration constituting a sound. Other subjects, according to the author, recognized a difference between sounds that would be at least 1/161 of the vibrations. However, this author does not say that this difference was harder to discriminate between the lower tones, or easier in the higher tones, and I have never heard that a difference between the higher tones, which have a larger number of vibrations, was more easily perceived than between the lower tones. I therefore suspect that in hearing, too, what is discriminated is not the absolute difference between the vibrations of the tones, but the relative difference compared with the number of vibrations of the tones.

This observation has been confirmed in many senses, that people perceive relative rather than absolute differences when making a discrimination. I have repeatedly felt driven to investigate the cause of this phenomenon, and I hope that when it is sufficiently well understood we will be in a position to make a better judgement about the nature of the senses. I will mention here only that I have, by this observation, refuted something that could quite plausibly be believed concerning the method by which we compare two impressions on the senses.

For instance, it would not be an inherently incongruous idea that we compare the images of two lines pictured on the retina in the same way as two objects held by hand. We compare the length of such things most accurately when one of them is superimposed on the other, so that we recognize both their similarity and their difference. The same thing could also occur in the eyes. A line might make an impression in the centre of the retina that persisted for a time while the eye was turned away: if the eye was turned upon another line so that its image fell on the same place, the length of this image could be compared with the impression of the line seen earlier, and their degree of similarity could then be appreciated. My observation, however, is inconsistent with this hypothesis. If the above method were the case, we should be seeing the absolute difference. For if a line that is equal to an *elbow* is superimposed on one that exceeds an *elbow* by one *thumb*, this *thumb* is seen as distinctly as if you place a line that is equal to a *foot* on a line that exceeds a *foot* by one *thumb*. Therefore, if one line is superimposed on another in the retina, the disparity would be distinguished as easily in the former case (where it is equal to 1/25) as in the latter (where it is equal to 1/12)*. [175] On the contrary, the visual experiment shows that we recognize a difference in length between lines placed next to each other more easily when the lines are 13 *thumbs* long and a *foot* (12 *thumbs*) than when they are 25 *thumbs* long and an *elbow* (24 *thumbs*).

It would not be patently absurd if someone claimed that the lengths of two lines were compared by vision and measured in such a way that the mind unconsciously counts the nerve endings in the retina that are touched by the impressions of both lines. My observations, however, are also inconsistent with this hypothesis; for if this were so, the absolute difference between the lengths of those lines would be seen, and in no way would the difference be perceived any better if it were larger in relation to the length of the lines.

The senses are easily fatigued and tired by continuous work, so that we no longer perceive differences that were obvious at another time.

This is the most productive source of errors, and it causes the most trouble to anyone who wishes to measure tactile acuity with precision. Other investigations, as is usual in experiments, are made more certain by frequent measurement and accurate repetition; but these measurements, on the contrary, become uncertain and erroneous when the experiments are continued and repeated for a long time. For this reason, measurements should certainly be repeated, but at different times, so that we never embark on the task with fatigued senses. No one will be surprised, then, that the work extended over the space of many years in bringing this series of measurements to a conclusion. Nevertheless, my work needs to be supplemented and corrected by the observations of others.

---

* For consistency these fractions should be 1/25 and 1/13, or 1/24 and 1/12.

FOOTNOTES

1. It should be noted that we are giving here the distance between the inside edges of the points. If we were to measure the separation of the legs from the middle of their stumps, the distance would be longer by a third of a *line*.

2. This experiment was carried out both when touching the mid-line of the tongue, and with the points touching the middle of the left or right sides.

3. Perhaps the points seemed to be so widely separated because one point touched near the rib and the other near the crest of the hip bone.

4. Sensitivity in the back of the upper and lower arm scarcely differs from that in the front, though it is slightly more acute.

5. The sensitivity of the back of the lower arm is almost as perfect as the front, but is slightly less acute.

6. The palmar surface of the thumb and fifth finger seemed to have slightly lower acuity than the three middle fingers.

7. See FERD.AUG.HOLKE *Disquisitio de acie oculi dextri et sinistri in mille ducentis hominibus sexu, aetate et vitae ratione diversis examinata.* Diss. inaug. Lipsiae 1830 apud Leop. Voss. 4. (An investigation into the acuity of the right and left eye examined in 1200 subjects, varying in sex, age and mode of life. Inaugural dissertation, Leipzig 1830, published at Leop. Voss. 4.)

8. Müller uses these words: *Es entsteht keineswegs die Empfindung der Mittelfarbe z.B. des Grünen von Blau und Gelb, sondern nur das Dunkle und Helle der beiden Farben gleicht sich aus, so dass das Gesichtsfeld weder so hell ist, als wenn durch das gelbe, noch so dunkel, als wenn durch das blaue Glas allein gesehen wird. Aber das Gesichtsfeld wird abwechselnd blau und gelb gesehen.* (If we combine blue and yellow, we never have a sensation of an intermediary colour such as green. But the darkness and brightness of the two colours is balanced out, so that the visual field is neither as bright as when viewed through the yellow glass alone nor as dim as when viewed through the blue glass alone. The visual field, however, is seen as alternating between blue and yellow.)

9. Delezenne in *Recueil de travaux de la soc. des sciences de Lille 1827*. The report extracted from this work is contained in *Bull. univ. des sc. math. XI 275*, and in *Repertorium für Experimentalphysik von Fechner, Leipzig 1832. 8. Bd. 1. p.341.* (Translator's note: C.E.J. Delezenne, "Sur les valeurs numériques des notes de la gamme" *Recueil des travaux de la Société des Sciences de l'Agriculture et des Arts de Lille, 1827, pp.4-6.* appears in an English translation by Mollie D. Boring in Herrnstein, R.J. and Boring, E.G. *A source book in the history of psychology.* Cambridge, Mass.: Harvard University Press 1965, pp.62-64.)

DER TASTSINN

UND DAS GEMEINGEFUHL

ON THE SENSE OF TOUCH AND 'COMMON SENSIBILITY'

ON THE CONDITIONS UNDER WHICH SENSATIONS ARE REFERRED TO EXTERNAL OBJECTS

The exact and quantified investigation of the sense of touch and of the 'common sensibility' inherent in skin and muscles is of particular interest insofar as we have the opportunity, offered by no other sense organ, to devise a wide variety of experiments which are harmless and to carry out many different kinds of measurements; moreover, some of the findings we might obtain concerning the skin senses may be applied later to the visual and other senses, not to mention 'common sensibility'.

Since all perceptible impressions on our body consist of movements affecting the body and causing changes in our nerves, the object of our sensations should appear to us to reside always in those organs by which we experience those sensations. Indeed, this seems to apply to many of our sensations. If for instance we have a headache, eye-ache, ear-ache, tooth-ache, or other pains, we sense that a particular part of the body is indeed in pain. We believe that we have sensations in that part of the body where the nerves are affected. We do not differentiate that which affects us from the organs which are being affected: we only notice the changes in the sensory state of our various bodily regions. Once the surgeon's knife has penetrated the skin, it is no longer perceived as an object coming into contact with a particular region of the body; instead we feel pain in the injured region itself. [4]

In regions which are not sensory organs we have only this kind of sensation. However by means of our developed sense-organs, we also receive other sensations, by which objects remote from the sense-organs are believed to be perceived. For instance objects that we see at a certain distance, appear to be spatially separated from us; yet we are also certain that the power of our nerves docs not extend beyond the surface of the body, and that we are capable of seeing those objects only if light from them strikes the nervous tissue of the eye, and gives rise to a tiny image of those visible objects. But we are totally unaware of these stimulations of the nervous tissue of the eye, even when we try to concentrate on them as much as possible: in so doing we are not even aware of having directed our attention to a region of the the retina, but, rather, of having directed it, necessarily, to a visible object in space.

With every sensation, we must distinguish between the sensation itself and our interpretation of it. The sensations of light, and dark, and colour are pure sensations; but we are interpreting whenever we feel that light, dark or colour is situated either within or in front of us, possessing a shape, or whether it is stationary or moving. However the interpretation is so bound up with the sensation that it is inseparable from it and is assumed by us to be part of the sensation, whereas it is in fact the idea which we derive from the sensation. Yet not only correct, but also false interpretations may sometimes be so completely intermingled with the sensations on which they are based, that it becomes impossible to divorce them, even if one knows one's error and the source of the error. Everyone, even the astronomer, sees the rising and setting sun and the rising and setting moon as apparently having larger diameters than they do when they stand directly above in the sky. However it is known that this illusion is not due to the refraction of light in the atmosphere, which supposedly results in a larger image on the retina. In fact the visual angles, in both cases, have been measured and found to be exactly identical. [5] Rather, the illusion is based on a false but inescapable interpretation necessitated by circumstances and so closely linked with the sight of the rising moon and sun that we find it impossible to separate interpretation from the sensation. We believe that we directly perceive the rising sun and the rising moon as having greater diameters than when they stand high in the sky. However we are not even aware of the reason our sensations should lead to this false impression. It has to do with the fact that the rising sun and the rising moon appear to be more distant from us than they do when they are directly above. All bodies viewed at the same angle appear larger if they are taken to be further away, and *vice versa*. The reason we assume that heavenly bodies on the horizon are further away than when they stand high in the sky is related to the fact that the arch of the sky appears to be shaped, not like a hollow hemisphere, but like a small segment of a hollow sphere, or a very curved watchglass. One can convince oneself of this if one imagines that to the arch of sky above one be added a second arch of the same shape, except curved downwards: one will then notice that the two do not form a sphere, but a lens. If the sky does not appear to be a hemisphere but to be a small segment of a sphere, then it follows that the distance to the zenith appears to be less than that to the horizon. But this clearly raises the further question of why the sky should seem to look like a small segment of a sphere. Many distant objects whose sizes are known to us, project themselves onto the horizon. From this we come to appreciate that those parts of the sky which are close to the horizon are actually very distant: whereas we do not have such a reference point for judging the distance of the zenith. A further possible contributing factor might be that all objects appear to be hazier, the further away they are; and hence we are accustomed to assume that hazy-looking objects are far removed from us, so that the sun and moon appear to be hazier, the closer they are to the horizon.

But it is not only with vision that we have this experience of referring our impressions, not to the point at which our nerves are stimulated, but to remote regions of space, believing that we perceive bodies affecting us actually to be there. [6] The same thing also happens with the perception of pressure *via* the organs of touch. The hairs are completely insensitive threads of horn-like substance, which can be burned without our feeling any sensation; yet they can transmit, like a plummet, movements or pressures into those sensitive parts of skin from which they grow. If the beard, e.g. on the cheek, is touched lightly, we do not feel the pressure exerted on the hair to be within the skin at that region where the horny thread is rooted and stimulates the nerves; we actually believe that we feel the pressure at a distance from the skin corresponding to the touched parts of the hairs. The same observation can be made regarding the teeth. The hard parts of the teeth are insensitive: pieces may be filed off them without arousing pain. Only the nerve-rich skin surrounding the root of the tooth and penetrating into the alveolus of the jawbone, and the dental pulp filling the tiny hollow in the tooth, are sensitive. If we now introduce a small piece of wood between the teeth and touch it with them, we believe that we feel the wood between the teeth; we think that we feel the resistance as being actually on the surface of the tooth, when in fact we can feel nothing there because of the lack of nerves. We do not have the slightest feeling of pressure on the surface of the root, hidden in the alveolus; yet the pressure actually spreads to the nerve-rich skin surrounding the root, and stimulates the nerves there.

But it is not only to the surface of insensitive substances covering our skin that we refer the site of a sensed pressure; we may even refer the sensation to the end of a rod inserted between our finger-tip and a resisting object, e.g. a table-top. Fechner* has drawn my attention to the fact that in these circumstances we have the impression of feeling the pressure in two places, one

* G.T. Fechner (1801-1887) was a Professor of Physics at Leipzig from 1833 to 1839; at the time of publication of *Der Tastsinn* he was undergoing the 'crisis' of his life, from which he would recover to start his experiments in psychophysics in 1850.

where the rod touches our finger-tip, and the other where the lower end of the rod touches the table. It is as if we simultaneously received two sensations, separated by the length of the rod. [7] I have described the conditions under which such double sensation arises. But if one end of the rod is fixed rigidly onto the surface of the table, and we press heavily on the rod, or even grip it tightly, the second sensation disappears and only the original sensation, perceived as localized where the finger touches the wood, remains. If we were to fix the rod rigidly on the end of our finger, while the other end could move freely over the table-top, the sensation at the finger-tip end would then vanish, and we would apparently feel pressure only from the lower end of the pencil, where it touched the table. In fact our teeth are like rods, fixed rigidly at one end in the tooth-socket; and this is why we appear to feel no pressure at the site where the nerve-rich skin of the socket is touched, but instead feel the pressure on the free surface of the tooth. Only when a tooth is noticeably loosened and moves in its socket, and a rigid object is pressed against the tooth, do we - as I myself have experienced - feel two sensations, one on the surface of the root, the other on the surface of the crown.

This therefore gives us an opportunity to investigate more fully the circumstances which lead us to interpret our sensations in such a way that we assume the distant end of the rod to be touching a second resisting body, the latter itself being felt to be situated at a definite distance. We feel the contact of the rod with the table most clearly when we revolve the upper end of the rod, with our finger, in an arc about the lower end of the rod on the table. The rod offers a certain resistance at every position along a given line of direction, and as these directions all correspond to the radii of the arc in which we move our finger, we judge that, in every direction in which the rod is felt to offer resistance, there must be a resisting body. Because it is unmoving, this must be distinguished from the moving rod. The more the rod and the finger move in unison, the more clearly we imagine the rod to be touching the table; but the more the finger moves about on top of the rod, and the less the rod moves with our finger, the more clearly do we feel that our finger touches the upper end of the rod, and the less clearly do we feel contact with the table. [8] Now we can really gain some insight into how it is that what we think we feel is actually given to us by a judgment based on a comparison between several feelings and on our awareness of our own movements. Either there is a rational interpretation of our feelings with regard to all these circumstances which really is based on a judgment (i.e. a synthetic judgment) formed before we have constructed concepts which could be expressed verbally; or we have this understanding vicariously, i.e. our soul is influenced by some unknown cause, without insight into these relationships, to interpret feelings in line with these relationships as if by intellectual instinct. Similarly, the phenomenon that we localize a sound outside our head and not inside, at the point where it excites the auditory nerves, is also based on a synthesized judgment. Let us do the following experiment, for example. If we turn our head to a point at which one ear, say the right, is oriented towards a source of sound, and the other ear is directed away from the sound, we find that the sound seems to be much louder in the former ear than in the latter. But if we now turn our head while the sound continues as before, the strength of the sensation in the right ear gradually decreases while that in the left ear increases. Finally, when our face or the back of our head is directly oriented towards the source of sound, the strength of the sensation in the two ears is equal: and if we continue to turn our head, it becomes stronger in the left and weaker in the right ear, until the difference between the sensations reaches a maximum. This observation, that turning the head can change the strength of sensations in such a lawful manner leads us to believe that the source of the sound remains constant and in the same position, with our sensations increasing and decreasing only by virtue of our head-turning: it would seem therefore that the position of the sound-source changes relative to our ears according to our movements. [9] From this we deduce that the source of the sound, if it continues unaltered, cannot reside within us, but must exist externally to us: for otherwise, it would also move as we moved and therefore sound the same despite our movements. All these phenomena can be completely understood in detail if we assume that the sensation of sound increases, the more precisely the external opening of the ear is directed towards the sound. The hypothesis that a physical source of sound lies outside our body can be confirmed in many ways, for instance by moving the sound-source towards or away from us, and noticing how the sensations increase or decrease as we do so. We can similarly convince ourselves that the source of many smells lies in space external to us, and not at the point where the odorous particle makes contact with the mucous membrane of our nose. If it was not possible for us to increase or diminish smell-sensations by moving towards or away from the source of the smell, or to render them more distinct by increasing the air intake into our nose, that is, if we had to renounce the ability intentionally to increase or decrease our sensations of smell, then we would turn to seek the cause of smells within ourselves, just as we do with the sensations of pain, nausea, hunger or thirst. This shows itself in a particularly interesting manner in the perception of warmth. Our skin-temperature can be raised in two ways; internally, by an increased flow of heat when more warm blood is sent to the skin, or externally, by the application of heat to the skin. In both cases we feel our skin to be hotter. If the hot object conveying heat from outside is actually pressed against the skin, we are left in no doubt that the heat comes externally to us and we feel that the object pressed against us is itself hot. But if we are surrounded by radiant heat or lightly warmed air, it is much harder to decide whether this warmth is affecting us internally or externally. [10] But even when making judgements on these kinds of sensation we are led to evaluate them by the same considerations as applied to the circumstances already mentioned. If we ask somebody to close his eyes, and then bring a glowing round iron bar about 1/3 inch* in diameter, to about 1 or 2 inches from his face, so that he has a vertical object in front of his vertically oriented face: and then ask him to turn his head repeatedly to the right or the left: then our subject will certainly sense the position of the heated bar at a certain distance in front of his face. For while the head is rotated about the vertical axis, the bar sends its heat-rays most intensely upon that region of the face nearest it, with the region varying as the head is turned in one direction or another. If the source of the heat were in our skin, it would move along with the skin and maintain its position relative to it. We infer that there is a stationary heat-source of linear form, both from the fact that a certain vertical region of the skin feels more heated than its surround, and from the fact that when we turn our head, different regions of the skin are affected in an orderly fashion by the heat, which are reversed when we turn our head back.

So we find that there is no local sensibility for vision in the eye, for audition in the ear-labyrinth, or for olfaction in the smell-sensitive part of the nose. On the other hand, we imagine we can feel objects on the surface of our teeth or hair, whereas there is no doubt that the hard parts of the teeth and hair are totally insensitive. Furthermore, as Joh. Müller** has shown, a pressure on a nerve-trunk containing many touch-nerves can give rise to a pain localized not only at the site of the pressure, but also in fairly remote regions supplied by the stimulated nerve-branches. And finally, we know that there are diseases in which severe pain is felt in regions far removed from the brain and spinal cord, whereas the site of the nervous disorder is actually in the cord or brain. So

* Presumably Weber is referring to the Paris inch (or thumb) throughout the text. See note on p.22.

** J.P. Müller (1801-1858), Professor of Anatomy and Physiology at Berlin from 1833. He is best known for his 'doctrine of specific nerve energies'.

from all this, we must surmise that *a sensation in itself can tell us little about the site at which the nerves conveying the sensation were actually stimulated; and that all sensations are fundamentally conditions exciting our consciousness, the kind and quality of which may be varied.* [11] *These conditions do not reveal spatial relationships directly to consciousness but only indirectly, insofar as our soul is capable of representing and integrating those sensations, and is induced to do so by an innate predisposition or power.*

The distinction between feelings and their representation.

The way in which we go about interpreting our sensations is not completely governed by our own free will; we feel obliged, for some unknown reason, to order our sensations with respect to the categories of space, time, and number. Were our free will not bolstered by this need when interpreting sensations, we should doubtless never succeed in achieving sensory representations. These representations are therefore not the result of experience, but rather experience itself is made possible by our ability to assign our sensations to the appropriate categories of space, time, and number. The awareness that we do not achieve the integration of sensations by a free activity of our souls is forced on us by virtue of the fact that if we try to interpret them differently, we realise that we *must* interpret sensations in a given manner, and that we cannot alter a single detail of the interpretation so given. We can neither omit one of the three dimensions of space, nor add a fourth to them. We can 'think away' our bodily world, but we will try to 'think away' space and time in vain. If we may be allowed to interpret the concept of 'instinct' rather more widely than is usual, by using this word to apply to the unknown cause of each of the innate purposeful activities undetermined by the mind, and thus relate such activity to image-building or to the production of movements, then we may refer to this state of mind as an intellectual instinct. [12] It appears that animals are driven by this same unknown cause to order sensations in terms of space, time and number, even though they are not capable of becoming conscious of this activity in abstraction, and hence are unable to form the concepts of space, time, and number. But it must not be thought that they have only pure sensations: the more advanced animals give evidence enough that they do not think that sensations provided by their eyes are actually in their eyes; for example, a dog seizes meat thrown before him with his mouth. Nobody doubts that dogs, cats, and horses search for sounds and smells in the space external to them, rather than within themselves.

We must be careful therefore, not to confuse the following internal processes:

1. movements in bodies surrounding us, which are trans-

mitted into the material of our sense-organs;

2. movements in our nerve-fibres, caused by the above movements, not being of a different kind;

3. changes in our consciousness excited by the movements in our nerves, and which we call *sensations*;

4. the ordering of the sensations in categories of time, space, and number;

5. the abstract concepts of these and all other categories, along with *concepts* derived from combinations of these categories.

In order to obtain an image of a sensation, the attention must be directed to the sensation we desire to reproduce, whereas the sensation itself can arise even when we are directing our attention at full strength to some other object. Sensations which we have ordered in terms of space, time, and number are more easily preserved in memory, whereas pure sensations which have not been so ordered leave no persistent impression and are therefore difficult to associate. Everyone has had the experience of many objects catching the eyes, yet he sees only those to which his attention is directed, and when he is very busy, many a sound can strike his ear without his hearing it. [13] We may here ask whether these impressions did not penetrate to consciousness and therefore excited only nervous movements, without any change in consciousness. I am inclined to think that such impressions actually do cause a change in consciousness, but that they leave no trace behind and therefore vanish rapidly away from us. If it is strong enough, the nervous activity aroused by a stimulus on our body can outlast the stimulus itself, and we can therefore occasionally still have a sensory representation after the external stimulus has ceased. We can see over a wide area during a lightning flash, and read a few letters by an electric spark, despite the fact that both are only momentary.

While I am counting the strokes of a clock, I can still see the shape of a flame and feel the shape of an object held in my hand. It appears therefore that we are capable of perceiving various sensations simultaneously. However, this experiment is not sufficient to prove this, for it may well be that our attention in the periods between the strokes is directed to the flame and then again to the shape of the tactually perceptible object, this alternation of attention being so rapid and frequent that it appears to us that the three sensations are being observed simultaneously and uninterruptedly. Just how short a time is needed to detect a sensation is exemplified by well-practised proof-readers, who read the proofs to be corrected rapidly, yet see each letter accurately enough to be aware of any mistakes which occur. For vision, I can prove that the region of the retina with which we can clearly see is only about 1/3 Paris *lines* wide*. We must therefore move our eye from one part to another so that gradually each part of a larger object is built up on this small and very sensitive region of the retina. Despite this we think we are seeing objects simultaneously which in fact simply could not be represented at one moment of time on that part of the retina. Items represented successively are apparently perceived simultaneously: the more exact investigations of Bessel[1]** indicate that, on the contrary, a visual and an auditory sensation cannot be detected simultaneously. [14] In observations done with chronometers, astronomers twice judge the distance of a star from a thread stretched across the viewing field of a telescope. The star itself passes across the thread: the observer must estimate how far the star is from the thread at the first stroke of a clock which sounds before the star reaches the thread, and again how far it is from the thread at a second stroke, after the star has crossed the thread. It was found that the observations even of the most practiced observers differ quite noticeably from each other because, so Bessel asserts, one observer will hear the stroke first and then judge the distance, while another will first judge the distance of the star from the threads, then hear the stroke.

Bessel's assumption seems to be confirmed by the following observations made by myself, which seem to indicate that it is not possible to perceive two different auditory sensations simultaneously in their time interrelationships if one is presented to the right and the other to the left ear. If I take two watches, whose ticking differs slightly in rate, and hold them near one ear so that the sound is only heard *via* that ear and not by the other, then I can distinguish those times at which ticks from the two watches coincide from those times at which the ticks of one watch fall in between the ticks of the other: I can perceive them as a repeated rhythm. But if I hold one watch next to each ear, while indeed I can perceive that one ticks faster than the other, I cannot perceive this repeated rhythm, and the ticking of the two watches therefore gives quite a different impression from that in the first instance. It is for the same reason that we cannot both hear our heartbeats and feel our pulse-beats simultaneously.

---

* One Paris *line* = .225 cm. or about .100 in.

** F.W. Bessel (1784-1846), Professor of Astronomy at Königsberg from 1810 onwards. He is known for many astronomical observations, the establishment of the 'personal equation', and the discovery of mathematical functions still known as 'Bessel functions'.

ON THE REASONS FOR OUR BEING ABLE ONLY TO RELATE A NUMBER OF SENSATIONS TO EXTERNAL OBJECTS [15]

But it is not so much by way of every region of our body as by our sense-organs that we receive sensations which we can so process that we can distinguish the effective stimulus-objects from the sensitive body-areas, and perceive them as objects touching our organs or even separated from them by large regions of space. We can touch one hand with the other, or touch the tip of our tongue with our teeth, or, *vice versa*, feel our teeth with the tip of our tongue in the mouth: in so doing we perceive them as external objects. Similarly we would be able to feel several internal parts of our body being touched by neighbouring parts, pressing on them and by this moving them, and come to an awareness of their shape and position, if only those inner regions were supplied with apparatus for touch-sensations.

But our sense-organs are directed outwardly not inwardly, in order that the mind may receive impressions from the external world: it would become very confused if internal processes were persistently demanding its attention. One intestinal canal touches and rubs against another, the lungs rub against the skin of the pleura covering the chest cavity, muscles press and rub against each other: but we have no sensations of these. We have earlier remarked that we can voluntarily move the diaphragm, a large arch of muscular skin separating the abdominal cavity from the chest-cavity: and when we do so, we can press large and massive organs such as the stomach, liver, and spleen down heavily into the abdominal cavity, stretching the elastic walls of the abdomen, as happens during breath intake or effortful bowel-movements: nevertheless, even if we attend hard, we do not perceive the existence of the stomach, liver, or spleen and have no awareness that something is being opposed to the diaphragm.

In fact all we feel is that we are sometimes making a larger, sometimes a smaller effort, and simultaneously feel an increasing tension in the skin of the abdomen produced by the organs of the abdomen having been pressed down. [16] Yet the diaphragm is not insensitive. We can experience severe rheumatic pains there and also have a sensation of the degree of the effort we make to move it. But we do not feel the pressure it undergoes, and have no sensations therein by which we could form an image of its form and position. Many a person without anatomical knowledge erroneously believes that, during breath intake the diaphragm presses upwards against regions in the chest; we know, however that, on the contrary, it presses the organs downwards into the abdominal cavity.

Now why should it be that, for some areas only, one's capacity to sense can be so developed that we can perceive objects, while for other regions it is impossible, despite a great exertion of attention, so that we can feel only change in our own general state of sensitivity?

The reason is that the *latter regions are so made that neither the movements of our bodies, nor the movements of sensible objects, produce any sufficiently noticeable alterations of sensation.* The movements of our organs, and of sensible objects, can bring about a change of sensation in two ways - namely, by making the feeling stronger or weaker, and, because of the intrusion of movements into different parts of the sensitive organ, by being affected in a way we are capable of discriminating. In regions where the above does not apply we cannot succeed in having sensations such that objects can be perceived. For instance the diaphragm can indeed be moved intentionally, but the apparatus is lacking which would enable us to distinguish the various degrees of resistance (pressure) undergone during movements of varying intensity: also there is no way in which we can tell which part of the diaphragm is being subjected to pressure.

But if impressions impinging on neighbouring parts of an organ are to give rise to not just one homogeneous sensation but to several distinctive sensations, the sensitive region must be arranged in a specific fashion: moreover, there must be a specific apparatus which allows us to discriminate between weak and strong impressions in order that various degrees of intensity may be distinguished. [17

We may note here that the sense-organs on the same surface possess much more numerous nerve-fibres than do other areas; and we have reason to believe that the nerve-fibres of those sense-organs, which, apart from the specific sensations they provide, also are the site for a more accurate sense of localization, are so arrayed that the arrangement of the peripheral endings corresponds to a certain degree with the arrangement at the central end-sites. On the other hand the nerve-fibres in the trunk of these nerves are not arranged in any particular order and hence do not always take the same route. Only the voluntary muscles parallel the sense-organs in terms of richness of nervous supply, but here it is the motor rather than the sensory nerves which are so numerous. If one mentally eliminates all the muscle-nerves, all the touch-nerves in the skin, and all the other sensory nerves of vision, hearing, taste, and smell, then extraordinarly few nerves are left to supply the rest of the body. The more thickly compressed the fibres of the sensory nerves in the touch-organs, the more easily we can localize sensations to small portions of those organs. In the finger-tips and in the tip of the tongue, which possess dense regions of nerve-fibres, for example, impressions can be localized despite the small region concerned, whereas similar impressions cannot be so discriminated on the arm or back, where the nerve-fibres are less thickly congregated. Fairly large movements of massive bodies are necessary to excite sensations with the assistance of the touch-organs; much smaller movements of massive bodies stimulate a sensation of sound in the hearing-organs; and only in the case of the eye can light, caused by extremely weak and rapidly repetitive movements of the imponderable ether, excite a sensation. Yet these latter cannot set even the lightest body, such as a mote in a sunbeam, into motion. The weaker the movements necessary to evoke an observable impression, the denser the external endings of the sensory nerves. Between the ends of touch-nerves is much insensitive material; the nerve-fibres are extremely dense at the points where the auditory nerves end in the labyrinth of the auditory apparatus; but densest of all are the nerve endings at the most sensitive point of the retina, which is along the optical axis of the eye. [18]

From the above we may say that the impressions evoked in us by movement, and the ability to discriminate whether our organs are stationary or whether the objects to be perceived are stationary and our organs in motion, lead us to distinguish the subjective from the objective, or put another way, to recognize the organs mediating sensations as being entities spatially separated from the objects affecting the organs. *When, therefore, autonomous movements of our organs or movements of sensible objects evoke no ostensibly perceptible change in our sensations, we cannot succeed, even with the greatest attention, in discriminating those objects from our sensitive body-regions.*

A change in sensation may be evoked by such movements when otherwise identical impressions can be distinguished because they affect two different parts of the body. But it is necessary that the two parts do not owe their sensibility to one and the same elementary nerve-fibre. In order that this feeling of localization be more fully developed into a sense of place, it appears that a definite arrangement in the nervous system is requisite: this arrangement consists in a division of the sensitive organ into small adjacent regions (sensory circles) from each of which arises a specific nerve-fibre running to the brain in isolation from other nerve-fibres. At the same time one may speculate that the nerve-fibres proceeding from the separate areas are arranged in the brain in the same adjacent order as they were in the organ itself. This is indicated by many of the phenomena observed after one-sided paralysis evoked by an effusion of blood in the brain. We find this incomplete sense of localization not only in all the sense-organs but even in regions possessing only common sensibility. But a more fully developed sense of localization is found only in the vision and touch organs, and then only really fully in the organ of vision. Experiments and measurements which I shall describe in more detail below have proved that the sense of localization in the various parts of the skin is developed very diversely for the sense to be 50 times more accurate in the tip of the tongue than it is on the back or the middle of the upper arm and leg. [19] Even in the eye localization is more accurate at the point where the optical axis meets the retina than it is in parts more distant from the axis: the more removed the visual regions are from those with the finest localization, the less complete and more coarse is the sense of place. In the auditory apparatus, localization is so undeveloped that we can tell only whether a sound comes to the left or the right ear: we cannot tell if it stimulates the auditory nerves in the cochlea or in the vestibulum. The taste-organ is similarly an organ of touch, and as such has an acute localization-sense: but as a taste-organ it does not. The sense of place is also undeveloped in regions supplied by the sympathetic nerves e.g. the spleen, liver, kidneys, etc. Sensory localization therefore seems to be the more acute, the smaller the neighbouring regions of the sense-organ provided with specific nerve-fibres, and the more densely packed the peripheral endings of those nerve-fibres running in an insulated fashion to the brain. In the case of the eye and of the organs of touch, which alone are the site of a well developed localization-sense, we can also tell most unambiguously that the objects being sensed are spatially separate from ourselves. After we have built up images of movement from these senses, and, furthermore, have become conscious of these movements we ourselves bring about, we have learned voluntarily to move the auditory and olfactory organs in such a way as to distinguish sounds and smells as also being objects separated from ourselves. If we had grown up in a stationary position like oysters, unable to intensify or attenuate our sensations of smell by moving towards or away from the source of the smell, or unable to regulate the sensation of smell by breathing in or alternatively prevent the sensation by not breathing in, we would consider smells as being merely changes in our state of general sensitivity, and not as external objects.

Sensations received by way of vision and touch are *sharply defined in space*: we can clearly discriminate two impressions in the skin even if they are presented to two nearby points on the skin and are similar to each other. [20] Simply by touching our skin with our finger, rubbing it often on the skin, we see on the one hand the layout of the small skin-regions in which we can discriminate the impressions, and learn which of the regions are adjacent and which are spatially separated; on the other hand we become aware of the effort of will required in order to move the finger in such a way that it touches the same regions of the skin in the same sequence. In this way we learn to move the finger intentionally; we feel its course as described on the skin. By the localization-sense in our skin we learn to recognise movements of our limbs, and, by movements of the limbs dependent on our will, we come to know the lay-out of the skin and to orient ourselves thereto. Both abilities, at first quite limited, are developed in a reciprocal way. After we have learnt the positions of the various small regions of the skin, we are enabled to recognize the shape of a body touching the hollow of our palm, without moving our touch-organs: for example, we can recognize the circular cross-section of a cylindrical metal tube, and as well the four- or three-sided cross-section of a four- or three-sided metal tube. From the location of the region of the skin pressed on, we can infer the shape of the stimulating region of the object touching us: from the pressure-contours* of our skin we can work out the contour of the object pressing on it. As the localization-apparatus in the middle of the retina is more than a 100 times finer than that of the most sensitive regions of the skin, we learn at an early age to deliberately move the eye and know the position of the most densely packed sensitive points. We can place the eye so that the objects to which we are attentive stimulate the middle of the retina, where the sense of localization is most acute and where objects are most clearly defined. From this it follows directly that the eyes are so devised that the extrapolated optical axes converge on the object being fixated. We learn to distinguish between the case where images on the retina move because our eyes move, and that in which the visible objects are moving, our eyes remaining stationary. [21] From the light-contour on the retina of the eye we can infer the contour of the light-emitting body.

The sensations provided by the sense-organs are however further distinguished by the fact that *they are very clearly circumscribed in time, i.e. they arise instantaneously with the stimulation of the sense-organ and last no longer than the stimulus itself*. This is not the case with many other sensations. All pains last longer than the stimulation causing them; and sensations in the nose giving rise to sneezing, sensations in the lips arising from a light touch, and shudders elicited by a light stroking of the exposed back with the tip of a feather, all last longer than the stimulation period itself. Moreover, many of these sensations do not arise instantaneously upon stimulation.

The organ of hearing excels all the above-mentioned sense-organs insofar as the sensations it gives us are very sharply defined in time. But the eye and the sense of touch offer much in this respect, too. One can easily see how much less effective the skin and the eye would be as regards providing images of movement if an impression

---

* *Figur* has been translated here as 'contour'

made at a sensitive point persisted for some time, so that successive impressions made at the same point appeared as simultaneous. This is the case with the eye when a glowing coal seen in the dark moves in a circle; in fact we do not see the movement of the coal, but a stationary glowing orb.

The sensations provided by the sense-organs are furthere characterized by the following: *Very faint impressions thereon, not in the least harmful, are nevertheless sensed quite clearly, so that we can definitely distinguish different degrees of sensation and even measure them and perceive countless qualitative differences.* Think how many countless variations in the quality and intensity of colours, tones, tastes, and smells, we can perceive! It is thus that we are enabled to discriminate small changes in the intensity of sensations arising when we approach or move away from an object or turn our sense-organs towards or away from it. [22]

We can even grade general sensations, e.g. pains; but how undeveloped we are at this, compared with the numerous degrees of temperature or levels of pressure which we can observe and to some extent measure with the sense of touch. If a warm body excites no pain, we can clearly distinguish, as we shall show below, a temperature difference of only 0.3 or 0.2 $^{\circ}$C; but if the temperature of the warm body does excite pain, we can no longer achieve such a fine discrimination. We cannot even make coarse discriminations. But pains can be distinguished qualitatively, and people have spoken of burning, pressing, boring, cutting, and many other types of pain. But it is very doubtful whether qualitatively different types of pain exist: these various differences probably depend on variations in the strength, extent, and duration of the pains. Tooth-aches caused by cold, warmth and pressure applied to the exposed pulp do not differ qualitatively. It is the same pain. In the various types of pain and particularly of common sensibility, a good deal depends on whether the site of the pain is extensive or limited in area, whether the feeling arises alternately at several distinguishable sites, whether it is very brief and repetitive, (e.g. the prickling sensation when limbs 'go to sleep'), or whether it is persistent and gradually increasing or decreasing in intensity. But so much is certain, that qualitative differences between pains and other feelings of general sensation are much less numerous than feelings from the special senses.

It is very important that impressions exciting sensations are not only very weak in themselves, but also that the sense-organs are specifically provided with devices for preventing the impressions from exceeding a certain level.

One effect of the intensity of many feelings provided by common sensibility is that the mind is prevented from calmly contemplating them in the manner necessary for the sensation to be referable to objects. Instead, the attention of the mind is driven by the pains to its own state of suffering, and its own body. [23] The effect of this is that the sensations excite not so much the cognitive faculties as the faculty of desire, so that we are driven to avoid the pain by instinctive or intentional movements.

There are several conditions under which we cannot refer sensations to external objects, and therefore shall count them as sensations of 'common sensibility'.

(1) When the sensations are not aroused by external objects, but by inner events in our organs, e.g. changes in substance in the muscles upon fatigue, changes in nourishment to regions, perhaps by an exaggerated blood-flow, as after inflammation, or because poison gets into the blood and comes into contact with nerves in the blood-vessels in various body regions, or because bodily changes occur consequent upon malnutrition or thirst;

(2) When the impressions are indeed occasioned by external sources in organs which do not have an organization suitable for the perception of spatial differences in impressions, or in degree and quality of the sensations;

(3) When the impressions are indeed occasioned by external sources, and which affect the appropriate sense-organs, but are so intense that they not only stimulate the nerve-endings, but also their trunks: pain is evoked, the faculty of desire is excited in the mind, and calm reflection is prevented.

(4) When impressions, made anywhere on the nerves, are not transferred as is customary to a certain region of the brain, but from the excited area are transferred to other areas of the body; but they are also transferred to the brain, and new sensations thereby evoked as a by-product, e.g. when we shiver if the point of a feather is stroked across the naked back, or feel an itch if a small body lightly touches our upper lip.

ARRANGEMENTS AT THE PERIPHERAL ENDINGS OF THE SENSORY NERVES FOR RECEIVING IMPRESSIONS ASSOCIATED WITH SENSATIONS [24]

In order that various naturally existing movements may give rise to specific sensations, and in order that many degrees of intensity of sensation can be clearly distinguished and measured, there must, as the foregoing indicated, be specialized sense-organs at the peripheral end of the nerves; these will be set into activity by the various movements corresponding to various degrees of intensity, and be so related to the nerves that by virtue of the movement into which they are put they cause a corresponding change in the nerves. Many natural movements are actually not sensed because the appropriate sense-organs are lacking at the peripheral end of the nerves. For example we are continuously exposed to magnetic influences from the earth which undergo daily variations. Yet magnetic storms pass by over us without our having the slightest awareness thereof. Electric storms would also pass by unnoticed were it not that some of the other electrical phenomena were perceptible to the eye and the ear. It would be quite otherwise were it the case that magnetisable bodies, when introduced to the peripheral ends of several nerves, were set in motion by the earth's magnetism and thereby caused changes in the nerves. We would then be able to detect the cardinal points of the compass by way of feelings. The sense-organs at the peripheral endings of the optic nerve are so arranged that vibrations due to light are concentrated and give them the appropriate orientation and bring about a change in the peripheral endings of the optic nerve, enabling us to sense light and colour. But these may only be affected by *transverse* and not by longitudinal vibrations of the light-ether: they are not so devised that sound-waves penetrating the eyeball shall likewise evoke a change in the optic nerve of a kind arousing a sensation. For if I apply the end of a sounding tuning-fork handle to the eye-ball, the vibrations must be transmitted throughout the whole eye-ball. [25] True, one does feel the vibrations on the skin of the eyelids, but the nervous tissue of the eye and the optic nerves are not thereby affected in such a way as to evoke a sensation of light, sound, touch, or even pain. And this seems to be true of all the sense-organs: each is so fashioned that it only reacts to a certain class of movements. That the sensory apparatus at the nerve-ends actually does play an important part in determining the end result can be seen from the fact that the touch nerves, if they did not have appropriate sense-organs, would be incapable of giving rise to sensations of pressure, warmth and cold: nor would the auditory nerves provide us with sound-sensations if they were not equipped with appropriate sense-organs at the periphery. Nobody would imagine that he would have sensations of touch and smell if gustatory or olfactory stimuli were brought actually directly into contact with the trunk of the taste or smell nerves; and we would have every right to doubt whether concentrated light focussed upon the freshly cross-sectioned surface of the optic nerve of a living animal would give rise to light-sensations. We do not have the opportunity to carry out such an experiment on living human beings. But for the sense of touch, we do have this opportunity. Hot or cold objects may be introduced so deeply into the body that they will make contact with the trunks of the touch-nerves lying superficially under the skin. Moreover experiments may be carried out with parts of the skin in which the touch-nerves have been destroyed by the intense and at the same time extensive application of heat. The experiments I shall describe in more detail below show that in the two cases the particular sensation of heat and cold are non-existent, if hot and cold objects directly stimulate the trunk of the touch-nerves. When I put the point of my elbow into ice-cold water, as in a mixture of snow and water, I feel, by way of the nerves ending in that submerged part of the skin, a sensation of cold. But after about 16 seconds the cold penetrates to the more deep-lying ulnar nerve, which contains a large number of closely-packed touch-nerves; and it affects this *directly, without the apparatus at the end of the touch-nerves being directly influenced*. [26] At this point we would suspect that from the moment the cold reaches the fibres of the ulnar nerve, the sensation would become more intense, because many more nerve-fibres are being stimulated simultaneously than previously. But this is not actually what happens: rather, from the moment the nerve trunk is directly affected by the cold, we feel a pain, which bears no similarity to the sensation of cold and which is not limited to the stimulated regions but seems to spread to include parts of the lower arm and hand. If the state persists longer, those fingers and parts of the hand innervated by the ulnar nerve become insensitive. It seemed to me desirable to set up an experiment using weaker degrees of cold which did not cause pain, and in such a way that the skin in which the sense-organs lay was not simultaneously affected. This could only be done by using a cold enema. If the rectum and flexura iliaca are filled with cold water and distended, the great anterior branches of the sacral nerve and the colon sinistrum lie so close to many of the dermal branches of the nerves of the loins that these nerves must undergo a considerable change in temperature.

This experiment was carried out with two good observers and myself. One of them had received an enema of about 21 ounces or 360 grams of water at a temperature of 15°R (18.7^{o}C). The water excited a strong feeling of cold in the anus upon its introduction and later expulsion. But within the abdomen or pelvic cavity the observer had no sensation of cold even upon the introduction, in a second experiment, of water at 6°R (7.5^{o}C). The same was felt by the second observer when water at room temperature was introduced. I myself had an enema of about 14 ounces (420 grams) of water at a temperature of 15°R (18.7^{o}C): as this was introduced, I felt a strong sensation of cold in the anus, and thought that I felt some movement in the abdomen as the bowel filled: there seemed to me a very faint, almost unnoticeable sensation of cold, which seemed gradually to spread up towards the middle of the abdomen. [27] But when, in a second experiment, I took the same amount of even colder water at $6^{\circ}R$ ($7.5^{\circ}C$), I had no clear sensation of cold, though I thought I had a faint feeling which I interpreted to have arisen from the intrusion of the water into the bowel. After some time had passed I thought I perceived a faint coldness, more in the region of the anterior abdominal wall than in the back. As this trace of cold might have penetrated from the colon (where it lay against the abdominal wall) out towards the external skin and thereby affected the ends of touch-nerves via the touch-organs of the skin, confirmation or refutation of this suggestion was attempted by placing a thermometer on that region of the abdominal wall at the point where it makes interior contact with the colon, the thermometer being covered with pieces of cloth. Over a long period it rose only to $27^{\circ}R$ ($33.7^{\circ}C$) whereas on the following day placed on the same site, it rose up to $28^{\circ}R$ ($35^{\circ}C$). So it really did seem that the skin at that part of the abdominal wall had been cooled by $1^{\circ}R$.

So we can be certain that if the large nerve-trunks in question, which contain countless nerve-fibres, were capable of transmitting the impression of cold independently of the touch-organs, and giving us a sensation of cold, then an intense sense of cold ought to have been felt. The above experiments therefore confirm the hypothesis that cold, when it acts directly on nerve-trunks, does not give rise to a sensation of cold. If I simultaneously and equally press both the skin and the ulnar nerve at the condylis internus ossis brachii, I feel a pressure by virtue of the touch nerves penetrating the skin at the elbow: but by way of the trunk of the ulnar nerve I feel either nothing or, if the pressure reaches a certain intensity, pain. The pain has not the least similarity to the sensation of pressure, but rather, like tooth-ache, is a peculiar type of nervous pain which seems to extend on the volar side of the ulna down to the hand, even as far as the little finger. *Thus the sensation of pressure and the discrimination of so many varied levels thereof only appear possible if the pressure works directly first on the touch-organs and thereby the ends of the touch-nerves, but not if the touch-nerves themselves be directly pressed.* [28] If things were otherwise, we should often have occasion to feel lively sensations of pressure: for it quite frequently happens that a nerve-trunk is accidentally pressed upon quite heavily. If intense pressure does not begin suddenly, we do not at once feel pain, but the limb 'goes to sleep'.

Below I shall describe in more detail experiments I have carried out with patients whose skin had been extensively destroyed by severe burns and the resultant suppuration, but was partly healed. *These led to the conclusion that patients with skin-regions in which the touch-*

receptors had been destroyed and had not yet been fully regenerated, could not distinguish between warmth and cold. It is a well known fact that large scars which have not yet achieved the texture and colour of the skin are incapable of touch-sensations, yet nevertheless give slight pain under certain circumstances which would normally not arouse pain in healthy skin. Some of these patients also confirmed this. One complained of scar-pains occasionally arising from changes in the weather. These phenomena may be explained by the fact that in order to feel heat and cold it is necessary that the stretchings and contractions invoked by heat and cold operate chiefly on the microscopic touch-organs frankly not yet known to us but which lie in the thick outer skin. From there the endings of the touch-nerves themselves are affected. If heat or cold are applied directly to the nerves, there is either no sensation, or there is pain. In the smell organ a small change affecting the smooth epithelium of the mucous membrane can rapidly lead to the loss of the sense of smell.

I have shown in experiments that if a man lies on his back so that the nostrils are directed upwards, it is possible to pour water into the nose through one nostril and thereby fill up both nasal cavities without the water entering the throat. [29] In this circumstance, according to Dzondi*, the arcus pharyngo-palatinus appears to contract by way of its enclosed muscles, thereby shutting off the exit from the upper part of the throat, behind the choanis narium, to the middle region of the throat. Thus the water, after it has filled the upper part of the throat, enters the choana narium of the other nasal cavity until both cavities are filled and the water runs out of both nostrils. Whether or not the water is at blood-temperature, the above procedure always has the effect of abolishing the sense of smell up to ½ minute or more after removal of the water. Even eau de cologne, pure acetic acid, and ammonia cannot be smelled. I would explain this remarkable finding as follows: the ciliated epithelial cells play an incontrovertible role in olfaction. The cells of the columnar epithelium have, according to my experiments, an unusual capacity of absorbing water but lose it some time after they have come into contact with pure water and been filled with it. Because of this they will doubtless be incapable, for some time, of effecting any absorption of the kind necessary for the olfactory material to act on the nerves.

---

* K.H.D. Dzondi (1770-1835), practised medicine in Halle and Greifswald. He wrote on many illnesses.

Exact evidence is still not available as to whether, if the vitreous humour escape from the eye-ball, or the fluid escape from the bony labyrinth (if the stirrup here be torn from the oval window), the ability to sense light or sound is momentarily lost; we only know that both injuries entail blindness and deafness respectively. But if the oval window is opened up as described, the hearing-nerve itself is quite uninjured. As sound is known to be transmitted in two ways to the auditory nerves, namely by the air in the entrance to the ear, or by bone-conduction in the head, it must be concluded that this kind of deafness must arise from the fact that the outflow of the labyrinthian fluid disrupts one of the arrangements necessary for the transmission of sound-impressions to the auditory nerves. The eardrum can be destroyed, and the hammer can even be dislodged from its position without deafness ensuing, for these accessories only perfect audition. One can still hear through the head-bones. If one sings or speaks and at the same time places a hand on the top of the head, the skull can be felt to vibrate on account of the sound waves. [30] Sound therefore penetrates in several ways through the bones to the auditory nerve and stimulates it, but the sound-waves do not evoke any sensation of sound without the mediation of the necessary accessory apparatus.

FURTHER REMARKS ON THE CHANGES PRODUCED IN SENSORY NERVES

Whether or not a certain accessory apparatus is required, it appears necessary that the impressions received by the nerves be transmitted by the nerve-fibres to the central nervous system of the organism before they become the objects of consciousness. By this we do not mean that the brain only is the seat of the soul and that the changes arising in the nerves exert no influence on consciousness: we mean only that we do not become consciously aware of impressions unless they have been transmitted to the brain. It is obvious that, by this transmission we are not referring to movements such as those of pressure, warmth, sound or smell. The physical characteristics of the nerves forbid such an assumption. We must rather assume that the highly varied types of influence excite characteristic kinds of movement in the nerves, capable of considerable modification and of which we become aware as sensations of warmth, pressure, light, sound, etc. It is probably that the impressions of warmth, cold, and pressure are transmitted to the brain *via* the same fibres of the tactile nerves, even though the sensations evoked by these impressions are different. In the same way sensations of light, sound, and smell may be excited by the same class of movements, for perhaps these movements need only vary in their rapidity or slowness for sensations of highly variable kinds to be aroused.

We cannot then ascribe a specific mode of conduction

to specific types of sensory nerves. There are no satisfactory grounds for assuming that the nerves of any particular sense possess a characteristic mode of conduction; rather, we should suspect that the process whereby such conduction is achieved is fundamentally similar, not merely in all the sensory nerves, but also in the organism's motor nerves. [31] For the fibres of the motor nerves and of touch-nerves cannot be distinguished under the microscope, and the substances of the various nerves, apart from the membranes enclosing them, also present no outstanding differences. But we should have expected such differences did the mode of conduction of impressions vary between different types of nerves. Moreover these same effects (mechanical injury by hitting or piercing, chemical injuries by heat or corrosion, and electric shock) which excite activity in the touch-nerves also excite it in the motor nerves should the latter's trunks be affected. Perhaps such activity only evokes muscle-movements or pain because the peripheral endings of the touch-nerves are not in contact with the muscles, nor the central ends of the muscle-nerves with those regions of the brain responsible for the translation of aroused nervous activity into consciousness. The same influences which interrupt the conductive capacity of the motor nerves also interrupt it in the touch-nerves. Examples are the sectioning, crushing, or stretching of nerves, and, finally, as I shall show in the experiments to be described shortly, appropriate heating or cooling of the nerves of warm-blooded animals. So if there is no good reason for assuming that the motor and touch nerves have modes of conduction based on differing processes in the nerves, there is also no good reason for believing that such processes differ between the different types of sensory nerves: for the fact that many sensory nerves have thin fibres with thin coverings, and others have thick fibres with thick coverings, is insufficient ground for convincing us of the latter. It would be easier to explain the difference between Magendie's findings that damage to the olfactory, visual and auditory nerves evokes no pain and Purkinje's observations that a blow to the eye gives rise to an impression of fire while pressure gives the sensation of alternating sequences of lights and colours, in terms of differences in the activity of the brain-regions at which these nerves terminate, rather than in terms of differences in mode of conduction between the nerves themselves. [32]

If nerves are pressed upon or heated or cooled to a certain degree, their conduction may be weakened or even abolished. We know that cutting through a nerve-trunk first of all robs us of the ability to voluntarily move those muscles receiving their nerves only from the peripheral portion of the sectioned nerves, and which consequently are no longer connected by nerve-fibres to the brain; and secondly, that we are prevented from sensing any impressions made to the region supplied by fibres from the peripheral portions of the sectioned nerves. The same result is obtained if a nerve is so compressed by pulling a thread tightly round it that a persistent change, e.g. crushing, ensues. Galen[2] relates that he was able to tie a loose thick thread (κροκαις ισχυραις) or a woolen thread (νημασιν εξεριων) about the nerves of a pig, tie a loop in it, then pull the knot so tightly round the nerve that the animal suddenly lost its voice but, to the astonishment of the spectators, regained its voice as soon as the knot was loosened. But if the nerves were too tightly compressed by a linen cord, it was crushed if the cord were stiff, or cut through if it were thin. However Valsalva[3],* and Morganus** did not succeed when they tried to tie a cord round the peripheral regions of the vagus nerves in dogs and interrupt their action by pressure, or revive the action by removal of the cord. Fontana[4],*** notes that in order to produce an artificial paralysis of a muscle, the nerve had to be pressed together so forcefully that the whole strength of his thumb and forefinger scarcely sufficed to produce the requisite pressure, even though the nerves were exposed and very small animals, e.g. frogs, were used in his experiments. [33] If the nerves were covered with soft parts, e.g. the diaphragm-nerves of a young cat by the abdominal skin, the pressure had to be inordinately increased. Care should be taken in all these experiments, (according to Fontana), not to pulverise the nerves between the fingers or other crushing objects, as destruction of their network immediately leads to the inability to move the relevant muscle, without subsequent recovery of the ability. Fontana concluded from this that it was only rarely that one could succeed in a living animal in abolishing the functioning of the *peripheral portion* of a nerve after pressure exerted on its trunk. Nevertheless Joh. Müller[5] has shown experimentally that long-lasting pressure on the trunk of an arm or leg-nerve can give rise to feeling of prickling, 'pins and needles' and notably of 'going to sleep' in that region of the limb supplied by fibres from the peripheral portion of the pressed nerve-trunk. I myself have done a series of experiments upon complete or incomplete blocking of nervous conduction by pressure, heat and cold. If I evenly pressed my own ulnar nerve at the condylus internus ossis

---

* A.M. Valsalva (1666-1723), Italian anatomist known for his work on respiration and the ear

** Probably G.B. Morgagni (1682-1771), a student of Valsalva who became an eminent pathologist

*** F. Fontana (1730-1805), Italian anatomist and physiologist known for his work on cells and nerve sheaths.

brachii I felt the pressure at that spot owing to the presence of touch-nerves ending in the skin of the elbow. But in addition there was evoked a characteristic nervous pain, which was not restricted to the site on which pressure had been exerted, and which had nothing in common with the sensation of pressure, but was rather a characteristic pain like tooth-ache. It extended on the volar side of the ulna down to the wrist and even to the metacarpal bones of the fifth finger and beyond. [34] By a gentle but sustained pressure on certain parts of the arm I could, without feeling any particular pain at the region pressed, make those parts of the arm supplied by the pressed nerves go to sleep. Those parts supplied by the ulnar nerve and those parts supplied by the median nerve could even be distinguished in extent by observation of the regions which went to sleep after such pressure and the condition of numbness gradually decreased if the pressure on the nerve-trunk was lifted, and completely disappeared in a very short time. If the reader does not know the regions of the limbs which go to sleep following pressure on the ulnar or median nerves, or on both at once, he has only to observe his limbs when they go to sleep, in order to confirm these findings. But moreover these states of limb-numbness have several degrees. In the highest degree, it is possible neither to move those muscles supplied with branches from the inactivated nerve, nor to feel warmth, cold, or pressure. In this highest degree the insensitivity can be so extensive that one's own arm, if grasped with the other hand in the dark, feels like somebody else's arm. My brother, Eduard Weber, once observed this on himself. But before it reaches this degree, one observes states in which the limbs are only partially 'asleep'. Contact with the inactivated finger or palm gives rise to a feeling very different from that of normal touch. The feeling is not restricted to the stimulated region, but spreads up over a larger area of the numbed region. Nor does it cease at the time the contact is withdrawn but persists for a long time afterwards; and it changes in locus, alternating between various parts of the skin as if they were being lightly pricked with countless needles from inside and thereby evoking a feeling of quivering movement in those parts of the skin of the numbed limb. From the fact that in incompletely numbed limbs, feelings may be ascribed simultaneously to many parts of the skin, it follows that the extents and boundaries of the limbs appear to be more clearly felt when they are not being touched than is the case with limbs not in a state of 'sleep'. [35] Sometimes a subjective feeling of warmth is obtained from a numb hand, though never, so far as I know, a feeling of cold. But then how can we explain the fact that under these conditions the 'common sensibility' of the skin furnishes us with so many feelings, whilst the sense of touch is reduced and in part suppressed? And how is it that momentary stimulation of a numb finger can induce feelings for so long, by a kind of after-effect constantly changing in locus? I assume that in a pressure-induced state of semi-numbness not all the fibres of the touch-nerves contained in the ulnar or median nerve are incapable of conducting impressions, but that when only some fibres are in a suitable state to conduct, many others in the vicinity are not, and no touch-sensation can therefore arise. To this I may add the suggestion that the movement induced by contact with the nerves can contribute to the recuperation, for conduction-purposes, of many of the fibres which had been pressed and that this recuperation itself is associated with a feeling which we call prickling or 'pins and needles'. These remarks are also of interest to medical practice, for in the course of a one-sided paralysis, hemiplegia, a pressure on the nerves in the vicinity of their endings in the brain can give rise to phenomena similar to those observed when limbs go to sleep, namely restriction of the touch-sense, along with the evocation of many of the indications of 'common sensibility'.

These experiments on the influence of pressure on the ability of nerves to conduct may be related to my experimental findings, already referred to above, concerning the effect of cold on the ulnar nerve. The cold is produced by immersing the elbow for some time in a mixture of broken ice and water, and the effect seems to be similar to that of pressure. On immersion one first feels the contact of the cold medium, mediated by the nerves ending in the skin of the elbow (branches of the cutaneus internus minor). After about 16 seconds the nerve-trunk of the ulnar nerve, which is not covered by muscles in this region, but lies directly under the skin and fascia, appears to be attacked by the cold. [36] A characteristic pain is felt spreading down the volar side of the under-arm to the ulna, the wrist, the palm near the little finger, and the little finger itself. This pain is quite distinct from the sensation of cold and has no similarity to it. If one did not know that the arm was in cold water, and if the cold were not felt in the skin of the elbow, one could not guess that cold was the origin of this pain. If the cold persists, this pain noticeably increases up to a certain point in time, and it takes some effort of will to endure it. It feels as if the little finger were not so free to move as formerly; although it actually still can be moved, the finger seems to have gone to sleep. But gradually the pain subsides despite the fact that the cold affecting the elbow remains unchanged. The relatively higher temperature of a piece of a metal can only be slowly and faintly distinguished from that of a piece of ice, using the little finger. The fourth and even the third fingers, to judge by their feelings, seem to be also affected by the cold. Sensation in the thumb, on the contrary, seems quite unimpaired. After about 12 minutes had elapsed from the onset of cold, twitching began in the fifth and fourth fingers and was even observed in the muscles of the underarm and the hand.

I need scarcely emphasise that the ability of the arm-tissues to conduct heat is too slight to allow the assumption that cold is actually conducted from the elbow to the hand. The above phenomena depend entirely on the fact that the trunk of the ulnar nerve in the elbow has been cooled, and that we interpret the pain arising therefrom as if it were located in the ends of the nerve-fibres when actually the fibres were cooled much higher up, where they passed by the elbow.

I have shown in another series of experiments that the touch-nerves of the fingers, the tongue, the lips and other regions, if they are dipped for one or two minutes in warm water of temperature 41°R (51.2^{o}C) or 42°R (52.5^{o} C), lose, after some time, their ability to convey impressions of warmth and cold to us. [37] Even the abilities to touch and to detect pressure are blunted, although they do not disappear entirely provided the stimulation is only continued for a short time. During immersion pain is also felt but it is not so intense as to be unbearable. The pain then decreases, and the finger develops a condition which I am obliged to compare with that of going to sleep. The same experiences were obtained if the finger was dipped for a minute or longer in a mixture of broken ice and water, with the difference that the pain took longer to reach its peak, gradually increasing over two minutes.

Not only the touch-nerves, but also the taste-nerves lose their ability to convey sensations for some time if they are cooled or heated. If the tip of the tongue is dipped into a vessel filled with warm water at 41-42°R (51.2-52.5^{o}C) and is held there for half a minute or a minute, and is then brought immediately into contact with syrup or powdered sugar, nothing can be tasted; one also notes at the same time that the touch-sense of the tongue, superior in acuity over that of all other regions, is so incomplete, that the tip of the tongue is in a state of numbness. Only after six seconds or even more does its ability to sense return. The same experience results if the tongue is held for half a minute or a minute or longer in a mixture of broken ice and water. Here there is also a feeling of pain which is very similar to that excited by hot water. I have carried out these experiments with several persons, and the result has always been the same.

ENDINGS OF THE SENSORY NERVES IN PARTICULAR REGIONS OF THE BRAIN

Specific differences between sensations also appear to be dependent on the particular arrangement of those parts of the brain to which the various sensory nerves conduct. [38] Exactly what goes on in our nerves, brain, and mind when the movements evoked in our nerve-fibres produce changes in our consciousness and thereby arouse sensations, will always remain a puzzle. Materialists will either be inclined to assume that our mind has no independent existence, but rather that the activities we ascribe to the mind result entirely from movements of the mechanism of our body; or to assert that the mind itself is a movable entity, set into motion by those motions arising in the nerves. Personally I am more inclined to believe that the mind is still a totally uncomprehended source of power. The laws of reciprocity obtaining between bodies must themselves have a cause. As it is a fact that we can produce movement in our body by an exercise of will, and as we are conscious that we can initiate movements, we are obliged to assume that our mind (soul), by its will, can produce mutual interactions between the molecules of its body, and particularly of the nervous substance of the brain, possibly in the form of attraction and repulsion. If we realise that the mind does much without the slightest awareness thereof, and, particularly, can act without prior intention as a result of being forced thereto by Nature, we can easily imagine that the soul can unconsciously and persistently exert pressures of movement upon the molecules of its own organ and thereby maintain an equilibrium between them. For instance, it might possibly emphasise or reduce the mutual attraction and repulsion between the molecules, according to definite natural laws. So when the molecules of the organ of mind are displaced by the nerve's movements transmitted to it the activity of the mind itself would change, also according to pre-determined rules. The dim awareness by the mind of such rapidly repeated changes of its own activity is perhaps what we call sensation. However these speculations go beyond the bonds of experience, making it impossible to prove them by observation or experiment; I shall therefore dwell no longer on them, and will prove nothing from it.

Our ability to perceive is unquestionably dependent on the structure of the various parts of the brain to which the various sensory nerves go, whether we comprehend the movements caused by the sense-impressions in a particular way, or whether our mind is not aware of them at all. [39] A blow to the eye produces a sensation of light, and pressure on the eye, according to Purkinje's investigations, causes us to see light or coloured figures gradually changing and altering in form. Even an electric shock to the eye causes a sensation of light, although the shock does not give rise to any light visible from a distance to others, and on the other hand a similar electric or galvanic current causes no visual or other kind of sensation. If I turned my eye inwards, and held a cold metal body, e.g. the end of a large heavy key of temperature $0^{o}R$ or even $-4^{o}R$, against the eye-ball at the outer corner so that the cold must penetrate through to the nervous tissue, I felt neither a sensation of cold, nor light, nor darkness: I felt only a pain which had its origin not in the conjunctiva, but either in the ciliary nerves of the choroid or in the retina of the eye. According to Magendie's* experiments[6] mechanical damage of the nerve-tissue of the eye occasioned by contact with a hard object, by piercing with a needle, or by tearing caused no pain in mammals, amphibia or fish; and he claims to have also observed this anaesthesia in man, when the lens was pressed upon. But birds moved whenever he touched their nerve-tissue with the point of his instrument, and at the same time clearly narrowed their pupils. From this point on we shall be concerned in detail with 'common sensibility'. In the same way he thought he observed that, in animals, damage to the trunks of the olfactory and auditory nerves in the cranial cavity evoked no pain, while in the same animals in the same operation damage to the fifth pair was very painful. If it is confirmed that mechanical damage to these three nerves evokes no pain, it is probably because those parts of the brain with which they are connected are incapable of arousing pain: for direct damage to many parts of the brain is known to be painless. [40] But perhaps a particular organisation of the sheaths of the elementary fibres is necessary in order for mechanical damage to excite the same feeling, particularly pain, and perhaps this organisation only exists for elementary fibres surrounded by two layers, as in the touch and taste-nerves. This arrangement is not present in the visual and olfactory nerves, nor in the thin vessels of the brain. It seems to me that we should not overrate the influence of the central organs associated with the inner endings of the nerves upon the evocation of specific and varied sensations, nor under-estimate the influence of accessory organs at the peripheral endings of the nerves. I am still not convinced that each influence of itself is so large that electric current through the eye is sensed as light, through the ear as sound, through the tongue as taste, through the nose as smell, and through the skin as a blow, so that one and the same stimulus can induce the characteristic sensation corresponding to that of the sense in question. Nor am I

* F. Magendie (1783-1855) taught anatomy and medicine in Paris. Best known for being the co-discoverer, with Bell, of the distinction between the sensory and the motor nerves.

convinced that very different influences acting on the same sense can all cause a similar sensation, for the sensation of light can arise, for example, not only from the the influence of light but also from mechanical blows, electrical currents, and chemical influences from the blood acting on the nervous tissue, the optic nerves, and the central organ of vision. If this were the case the structure of the various central organs of these many senses would be so different that the contrasts would be striking, whereas in fact they are not. It is true that a wind can whip up a sea here, or whistle through a cleft or sound an Aeolian harp there, and yet at a third place set an anemometer into motion and thereby tracing out figures, registering its own motion by changes of direction; but in order for such a single cause to elicit so many varied effects, very different entities are required on which it can operate.

The resolution of this controversy is so important that it is necessary to devote particular consideration to the relevant facts.

If a man is struck in the face, it may be that he perceives this impact as a blow through his sense of touch, as a sound through his auditory sense, or as a flash through his visual sense. [41] But a blow to the tongue causes no sensation of taste, nor does pressure on the mucous membrane of the nose evoke a smell. The same blow causes many kinds of movements, compresses the skin, causes sound-waves to be carried through the air and the fixed parts of our heads, and even affects imponderable substances, for a blow can force heat and light from objects. So if we sense the pressure induced by the blow by touch, its sound-waves by the auditory organ, and the movements of imponderables invoked in nervous tissues by a blow on the eye as light, this must probably be in part put down to the action of accessory apparatus associated with the peripheral endings of the various various sensory nerves; and these must be such that a mechanical pressure has no effect on the optic nerves, but does on the touch nerves: sound-waves have no effect on the optic or touch nerves, but do on the auditory nerves: and vibrations of the ether have no effect on the touch or auditory nerves but do on the optic nerves, thereby eliciting a sensation of light even if they are too weak to elicit the same impression on other persons from a distance. This would explain how blows to the tongue or blows to the nasal mucous membrane do not arouse tastes or smells: the blows cause no chemical effects under the conditions prevailing.

It is assuredly the same when we come to study the effects of electricity on many sense-organs. If we approach a conductor charged with electricity, our face feels as if it were touching the threads of a spider's web. The leaping of electric sparks on our skin and the discharge of a Leyden jar or Voltaic pile through our limbs produces the sensation of a stab or a blow. But one of the effects of electricity is to make our hair bristle and stand on end; and as this applies to the fine hairs on the skin of the face, the former feeling may well have its origin in this movement of the hairs. [42] The spark of a Leyden jar forces the air apart and thereby evokes a sound, and can also be visibly perceived to penetrate fixed bodies and thereby induce mechanical effects: the touch-sense is therefore influenced by these phenomena. Volta[7],* made the following observation on hearing: a pile of 30 or 40 plates was closed through two blunt metal probes. He placed these in his ears and then brought the ends of these probes into contact with the pile. The moment he closed the circuit, he felt a vibration through the head, followed shortly by a sound or noise difficult to describe, but which he compared with the crackling made by a stiff dough being cooked. This persisted uninterruptedly so long as the circuit remained closed, but did not increase in loudness. Ritter,[8],** who has observed many things unconfirmed by others, says that he heard the tone of *g* when the circuit was closed. My brother, Eduard Weber, filled both entrances to his ears with water and introduced two metal rods, thereby closing a circuit and evoking, through the induction of two very large magnets, a strong current under particularly favourable circumstances. Every time he initiated the induction, he perceived a light appearing to jump slant-wise over the head, but heard no tone, in fact not even a sound. Heydenreich[9] describes how he put two thick lead wires deep in the external meati. But the bursts of humming and roaring sounds which he heard when he closed a pile with the wires were similar to those evoked when there was no pile in contact. It does not therefore appear that electric or galvanic currents act directly upon the auditory nerves to produce a sensation of sound. But they certainly do so indirectly, for if, for instance, the muscles of the ear ossicles are contracted, the ear-drum must be vibrated, giving rise to a sensation of sound, inaudible to others. As for *taste*, Volta found the following: two different types of metal brought into contact would evoke a very distinctive taste on the tip of the tongue: apparently the taste was sour when the tip of the tongue touched zinc, and alkaline when it was touched by the other metal. [43] In the

---

* A.G.A.A. Volta (1745-1827), Italian physicist who invented the electric battery.

** J.W. Ritter (1745-1810), pharmacist and physicist known for various discoveries in electricity.

latter case the taste was less marked but nonetheless sharp and unpleasant: and in both cases it lasted for several seconds, gradually increasing if the metals remained in true contact. Now the fact that the taste was perceived not merely during opening and closing of the circuit, but also while it remained closed, proves that the galvanic current is not acting directly on the taste-nerves to produce the sensation, for a uniform current arouses no sensation in the sensory nerves, and produces no muscle-movements by way of the motor nerves. But we do know that a uniform galvanic current can decompose the salt in the saliva, and that the liberated acids are attracted to the zinc pole and the liberated alkalis to the copper pole. If we touch the lower surface of the tongue, endowed with a very meagre sense of taste, with one of the contact wires, and touch the upper surface of the tongue, endowed with a very sensitive sense of taste, with the other, we will always taste that particular substance collecting on the pole in contact with the upper surface. Heydenreich* has confirmed this finding, also reported by Valentin** and others, in the following experiment. He took the wires from a Voltaic pile consisting of ten pairs of plates, but did not put them directly on his tongue; instead he first wrapped the wire of the positive pole with blue litmus paper moistened with distilled water, and the wire of the negative pole with red litmus paper. When he then felt a sour taste, the blue litmus was seen to turn paler at the same time. The alkaline constitution of the oral fluid prevented it from turning red. The red litmus on the other hand rapidly turned blue, the rapidity being much greater than when the galvanic pile was not connected. In the latter instance the weak alkaline activity of the oral fluids also turned the paper blue, but much more slowly and to a lesser extent. He did not confirm Pfaff's*** experiment as reported in Joh. Müller,[10] for when he took a tin beaker filled with lye in both hands, and dipped his tongue into the lye, he perceived an alkaline, rather than an acid taste. [44]

The *smell of phosphorus* evoked by frictional electricity in the smell-organ is well known. Volta tried in vain to evoke a sensation of smell by leading galvanic

---

* Possibly F.F. Heydenreich (1790-1872), mathematician and physicist of Tilsit.

** G.G. Valentin (1810-1883), physiologist in Berne. He is best known for his pioneering work on sensation and on the vascular system.

*** C.H.B. Pfaff (1773-1852), physician, physicist and chemist, best known for his work on the effects of electricity on animal tissues.

currents through the nostrils. On opening and closing the circuit, he only experienced a ripping pain varying in intensity, or a vibration varying in its extent. As regards the apparent smell of phosphorus, we know that Schönbein has discovered that it originates in the build-up under the influence of frictional electricity of a specialised substance, ozone; it is therefore certain that this smell can be ascribed to the direct activity of electricity upon the nerves.

It only remains to discuss the effects of electricity upon the eye, these being not so very surprising in view of the probability that light, heat and electricity themselves are phenomena based on movements of the light-ether. However, we must admit that more attention must be devoted to the visual phenomena evoked by a current passed into the eye or by pressure of fairly long duration, and perhaps by sectioning of the optic nerves. This also applies to the phantasms arising from pathological affection of the brain or of the retina.

Without the collaboration of the brain, or perhaps part of the spinal cord, no sensation enters consciousness, no memories are aroused, no muscles can be moved voluntarily, no thinking can be carried out, and impressions conveyed from the sensory nerves cannot even be transferred to the motor nerves directly via the nerves of the organic part of the nervous system, but only by way of the brain and spinal cord. Even if we do not have good reason for conclusively assigning the seat of mind to the brain and spinal cord, these centres of the nervous system are the media without which we could not become aware of the influences acting on the mind, and without which the mind could not apparently act upon the body. Consistent with this view of the importance of the central nervous system are the following facts. 45
[45]

(1) All animal nerves run together in the cord or brain, but on their way from the periphery to the cord and brain are not connected together by their nervous pulp. In amputated arms and legs and other limbs impressions made on the sensory nerves can only be transmitted if these are connected to at least a part of the spinal cord or brain.

(2) If a nerve is sectioned or its conductivity interrupted by some other means such as pressure, cooling, or heating, those regions supplied by nerve-fibres from the affected point are capable neither of sensation nor of voluntary movement. Damage to the sense-organs does not however prevent the formation of images of such sense-impressions as had previously been received from these organs.

(3) Many brain-injuries result in sudden death or

in complete insensitivity of the sense-organs; and many temporary influences of slight importance on the brain result in temporary unconsciousness, beginning with the patient's "seeing things go black before his eyes" despite the fact that his eyes have not in any way been affected. A haemorrhage in the visual cortex or striate cortex or other small region in this vicinity can produce a complete or partial disruption of the sense of touch, and a paralysis of certain muscles in parts of the body far removed from each other on one side of the body: and this despite the fact that the origin of the trauma has never affected these regions directly. In contrast, amputees think for years that they can feel sensations in limbs which no longer exist, being unable to rid themselves of this illusion by thought presumably because the remnants of the nerve-fibres can still produce impressions in the brain of the kind once proceeding from those limbs. [46]

(4) In cases of brain disease, patients think they have sensations of sights, sounds, touch, and even smells, when there is in fact no external object acting on the senses, and the sense-organs themselves are functioning normally: and in diseases of the brain and spinal cord patients claim to feel pain in healthy parts of the body.

(5) Many poisons, such as strychnine, which enter the bloodstream and thus penetrate throughout the body, causing tetanus and loss of sensation, only have these effects by virtue of their action on the spinal cord or brain. They do not act directly on the nerve-trunks: for a limb whose nerve-trunks are severed is not attacked by tetanus.

(6) My experiments have shown that when we have sensations we cannot directly recognise the site at which our nerves were acted upon by external objects; we have only learned to do this by the uniting of many experiences with such sensations.

On the sense of touch in particular.
Sensations of localization, pressure, and temperature

The sense of touch provides us with two characteristic kinds of sensation, *sensations of pressure* and *sensations of temperature*; but at the same time the touch-organ and its nerves are so devised that these two sensations are distinct from each other when they are evoked at two separate places on the skin. So we may describe the *sense of localization, pressure, and temperature* as three distinct capacities of the touch-sense. Compression or extension of the sensitive organs, or tension such as is caused by a weight pressing on the skin or the hair being pulled, give rise to sensations which we can concisely comprise under the name of pressure-sensations. [47] Temperature-sensations are either *positive*, i.e. *sensations of heat* due to raising of the temperature in our sensitive regions by the application of heat, or *negative*, i.e. *sensations of cold* due to the lowering of this temperature by the withdrawal of heat. Only those body-regions supplied with organs of touch can provide us with sensations of pressure and temperature. The inner parts not possessing touch-organs may be pressed, warmed, or cooled without evoking the slightest sensation of pressure, warmth, or cold. All sensations, other than the above, with which our sense-organs provide us, belong to common sensibility. Pains arising from pressure, warmth and cold should therefore never be confused with the sensations of pressure, warmth and cold. Moreover we must distinguish between the above pure sensations and the images they give rise to, particularly after the image of movement generally, or a consciousness of one's own movement specifically, has been evoked. Of primary importance in this context is the image of *resistance* offered to us by bodies when our own body moves: we shall say more on this below. The sensations of pressure and of warmth or cold are so different that it may be held doubtful whether the two should be seen as varying modifications of one and the same sensation. The tongue is the site of two senses, taste and touch; it may therefore be asked whether the skin is also the site of two sense, pressure and temperature. If the same microscopically small sense-organs at the end of the touch-nerves served both purposes, conveying the perception both of touch and its various gradations and of warmth and cold and their various gradations, the sensation of warmth and cold would have its origin in the perception of that pressure-change evoked because warmth expands substances, while cold reduces their volume; so we would need to assume only one sense, the sense of touch, in the skin. One would then expect that pushing and pulling in a specific direction on the skin would evoke a sensation of pushing and pulling, whereas in certain regions of the skin compression and extension in many directions would lead to a sensation of cold and warmth. [48] But if on the other hand, there were two kinds of organs in the skin, one being moved by pressure and thereby causing changes in its associated nerves, the other being moved by variations in temperature and thereby producing changes in its associated nerves, then one would have to accept that there was both a pressure-sense and a temperature-sense in the skin. The first of these assumptions seems to have more in its favour than the latter. I base this opinion upon observations I have made which appear to warrant the attention of physiologists: *cold bodies lying on the skin appear to be heavier, and warm bodies lighter, than they should be.* The sensation of cold appears to summate with that of pressure; that of warmth does not appear to do so, but rather may be considered to act like a negative pressure, thereby reducing the simultaneous sensation of pressure. Take two equal weights of exactly the same form, which can be conveniently stood one on top of the other. New thaler* in very good condition serve the purpose admirably. Let one be cooled below freezing-point, e.g. to -7° or -4°C, and the other heated to about +37° or + 38°C. Have an observer lie so that his head is completely supported, with the surface of the forehead horizontal; his eyes should be closed. Then lay a cold thaler on his forehead; remove it immediately: and lay two of the heated coins, one on top of the other, at the same site. Then remove these and quickly replace them with a cold one. When this is taken away, put the two warm thaler back, and repeat these actions until the observer is ready to judge which was heavier, the first or the second of the weights laid on his forehead. The observer will assert that both weights were equally heavy or even that the weight consisting of two warmed thalers was the lighter. This experiment, which I repeated on several persons with the same result, shows that the impression of cold markedly raises the sensation of pressure: for the cold weight is not only judged to be heavier when it is equal in weight to the other, but when it has only half the weight of the other. [49] It is obvious that if this experiment is to be successful, the observer must be completely supported and not allowed to rise; for then he would have an image of the heaviness of the weight not merely derived from the pressure it exerted but also from the exercise of his muscles. The latter concept would be necessary in order to lift the weights.

ONLY THE TOUCH-SENSE PROVIDES US WITH SENSATIONS OF PRESSURE AND TEMPERATURE

Physiologists have hitherto scarcely doubted that the inner organs which are not supplied with touch-organs, were nevertheless capable of conveying sensations of pressure, warmth, and cold. It seemed to me doubtful that these particular sensations could even be possible if the sense-organs necessary for their perception were lacking. In order to obtain some certainly on the matter, I invited my friend Dr. Günther, Professor of Surgery at Leipzig, to collaborate with me in some experiments in

---

* A German coin, no longer minted, worth 3 marks. See note on p. 66.

which we examined the ability of patients, in whom extensive and severe burns or other accidents had destroyed by festering a large portion of the skin, to discriminate between warm and cold metallic objects in contact with the surface of their wounds. One spatula was therefore placed in water of 7-10°R (8.7-12.5°C), and another in water of 36-40°R (45-50°C), until they attained these temperatures. When the warm and cold spatula were now brought in fairly rapid alteration against the surface of the wound, and patients were asked whether the object in contact was warm or cold, they gave a wrong answer as frequently as a right, even to the extent that a warm object was affirmed to be cold three times in a row. But if these experiments were carried out on uninjured regions of skin adjacent to the wounds, they could distinguish the temperatures easily and accurately. In one case the spatula was heated up more, and aroused a sensation of pain at the site of skin-destruction: this was not the case in the aforementioned experiments. Such injured or wounded portions of the skin, as is well known, are very sensitive: even very weak stimuli evoke pain. [50] Even changes of the weather can elicit sensations in such regions in a manner otherwise incomprehensible, yet despite this the ability to distinguish warmth and cold is not merely not raised, but actually impaired. Through Dr. Günther I also had the opportunity to observe, in one case, the insensitivity of the intestines to cold. Several intestinal coils had penetrated through an abdominal wound and were only covered by the extended abdominal skin. They were covered with a cloth soaked in water at room-temperature (September): the patient had not the slightest feeling of cold or pain and did not even feel pressure.

The same conclusion was drawn from experiments carried out by Steinhäuser[11] with a woman in whom, as a result of abdominal pregnancy, an abscess had formed in the lower trunk with an *anus artificialis* in the large intestine (this later healed). This woman, who was otherwise completely healthy and who could digest well, had an opening $1\frac{1}{2}$ inches in diameter through which the partly inverted intestine extended. Steinhäuser first covered the woman's eyes with a linen cloth, then touched the mucous membrane for some time with ice, and later, with iron so hot that it could scarcely be held in the hand. But the patient felt nothing of it. If the mucous membrane were pierced with a needle, she did not notice that she had been touched. These experiments were repeated often, always with the same results. Even when touched with lunar caustic*, or when a small piece of the membrane were cut off with scissors, the patient felt nothing.

---

* Fused crystals of silver nitrate, used to heal ulcers, etc.

When very hot or very cold fluids are swallowed, we can observe that the tongue, palate, and oesophagus are sensitive to touch. But from there on the sense disappears or is so incomplete that one doubts its existence. If the stomach be filled with warm or cold drinks, or the large intestine be filled with warm or cold fluids way of an enema, the membranes and muscles in those areas must receive the effects of the heat or cold within a few seconds. [51] This ought therefore to result immediately in an unmistakable sensation of heat or cold. But at temperatures not inducing pain in the skin, there are no such sensations. Admittedly, if the heat or the cold are so extreme as to be painful if applied to the skin, these fluids can also excite sensations in these internal regions, sensations of common sensibility, though rather weaker, but the sensations are not of heat or cold, nor is it possible to distinguish between different degrees of the sensations. In order to investigate this myself, I took a glass full of water, which I had left standing by the window in a frosty cold, and added enough snow to lower the temperature to one of 0°R. I then quickly drank the contents. I clearly felt this extreme cold in the oral cavity, the palate, and the throat. But I did not feel the gradual descent of the cold water down the oesophagus. In the region of the stomach, I did feel a sensation, which I took to be that of mild coldness, but as I could only perceive this in the region of the anterior stomach-wall, and not towards the back in the posterior region, I suspect that the sensation originated in a propagation of the cold to the skin of the abdomen in the vicinity of the stomach. I also carried out the converse experiment, drinking as rapidly as possible three cups of milk: the first had a temperature of $+56^{o}$R (70°C), the third $+50^{o}$R (62.5^{o}C), and the second had a temperature midway between those of the other two. I felt the heat in the mouth, the palate, and the throat, but not in the oesophagus. At the moment the swallowed portions came into the stomach, I had a feeling which lasted for some time: but it was not a clear feeling of heat, for I could have confused it with that of cold. Experiments carried out by me with several persons administered cold enemas have been already described on p. [26]. These persons also confirmed that interior regions not supplied with organs of touch are incapable of yielding sensations of heat or cold; and areas not provided with touch-organs cannot yield sensations of pressure either. This has already been indicated by the experiment with the diaphragm which may be confirmed by any person who has the opportunity to carry it out, and also by the observation that pressure exerted on the trunk of a touch-nerve evokes, not a sensation of pressure, but of pain. [52] It is the same when warmth, cold or pressure affect the exposed dental pulp: a nervous pain, similar for all three types of stimulation, is evoked, which has not the slightest resemblance to the sensations of warmth, cold, or pressure.

The mucous membrane of the nose is supplied with touch-organs only at the anterior entrances to the nose, and on the bottom and in the vicinity of the bottom of the nostril. For the higher regions supplied by the olfactory nerve, where the mucous membrane is covered with smooth epithelium, the sense of touch seems to be absent although the skin has a lively common sensibility. If therefore we breathe heavily in very cold air on a cold winter's day through the nose, the cold is felt at the entrance to the nose, on the bottom of the nostril, and on the upper surface of the velum palatinum, but not in the higher region. Similarly if a cold round smooth rod of iron be introduced into the nose, its coldness and pressure are felt at the entrance; when it reaches the higher regions it produces only tickling or pain, certainly no sensations of cold or pressure.[12]

THE ELEMENTARY FIBRES OF THE TOUCH-NERVES AND THEIR PERIPHERAL AND CENTRAL ENDINGS. [53]

The skin is both an organ of sensation and an organ of secretion. In it we can perceive *smaller organs* devoted to these purposes. The tactile papillae and hair-bulbs, and perhaps also the Pacinian corpuscles, constitute the apparatus for sensation: the skin glands belong to the secretory system.

The relations between the finer structure of the skin, as revealed by the microscope, and the sense of touch are still very little understood.

The *elementary fibres of the touch-nerves* cannot in outward appearance, be distinguished from those of the elementary fibres of the nerves of muscles; hence they have a diameter four times that of the olfactory or optic nerve, and about twice as large as that of the auditory nerve. But the taste nerves seem to have the same appearance and size as the touch-nerves. This difference in size is in part due to the thicker sheaths which protect the elementary fibres of the touch-nerves, taste-nerves and muscle-nerves from the pressure of surrounding parts, especially the muscles. As the olfactory, optic and auditory nerves are already well protected along their short paths, a less thick sheath is required for their elementary fibres. They are therefore delimited by simple boundaries, while the contours of the elementary fibres of the touch-nerves are doubled, as compared with the previously named three nerves. The touch-nerves are also much harder. To Bell*, we owe the discovery that

---

* Sir C. Bell (1774-1842), anatomist and surgeon. In his *Idea of a New Anatomy of the Brain* (1811) he claimed that a nerve fibre is either motor or sensory but not both.

there are specialised nerves for feeling and movement, and that the sensory nerves of the skin and other parts of the trunk and its extremities, when they reach the vicinity of the spinal cord, separate themselves from the muscle-nerves with which they had intermingled for the greater part of the route, and form the dorsal root of the spinal cord along with a special ganglion. [54] Magendie clarified this discovery through some most valuable experiments, and Joh. Müller's research has put it beyond all doubt. As R. Wagner* has observed, the sensory nerves may be distinguished from the motor nerves (and also the three sensory nerves mentioned earlier) by the fact that their elementary fibres, at the point where they pass through the spinal ganglion, are interrupted by a ganglionic swelling. It must be left for future work to decide the purpose of this intercalated ganglion-swelling into the elementary fibres of the sensory nerves. I have already pointed out[13] the great importance of Fontana's discovery, confirmed since by Prevost**, Dumas*** and Edwards, that the elementary fibres of the nerves, on their way to the brain, neither receive nor send out branches; they travel as simple, undivided fibres of fairly even thickness. I have shown that this result of microscopic anatomy may be supported by physiological experiments on the functions of the nerves. I said that the transmission of impressions appeared to be mediated only by the smallest nerve-fibres directly stimulated, and that the stimulation did not appear to be transferred from one nerve fibre to another because they were not connected together by any nervous tissue. This arrangement means that a specific site on the organs of touch is connected to a specific site in the brain by a single undivided fibre. It does not seem to matter what diversions the nerve-fibres should make en route, nor within what nerve-sheath they lay along with other nerve-fibres: but we may surmise that the order in which the nerve-fibres leave the skin corresponds to a second ordering, namely, that of their endings in the brain. Joh. Müller[14], in his outstanding investigations, has confirmed the fact that the elementary fibres neither receive

---

* R. Wagner (1805-1864), Professor of Physiology at Göttingen. Best known for his work on zoology and as editor of the Handbook of Physiology in which *Der Tastsinn* appears.

** P. Prevost (1751-1834), philosopher and physicist of Geneva.

*** C.L. Dumas (1765-1813), Professor of anatomy and physiology at Montpellier.

nor give off any branches, and has drawn most interesting conclusions therefrom. [55] On the question of where and how the touch-nerves end in the brain, we cannot yet give any definite information, but experiments we have made on persons with one-sided paralysis, hemiplegia, allow us to indentify the region of the brain which is the central organ for the tactile sense: *it is beyond the visual lobe and striate cortex, towards the circumvolutions of the cerebrum.*

In hemiplegia, Nature herself experiments for us. If there is vascular damage in the brain, it often causes pressure on a very small area of the brain, the lesion even being so restricted as to cause no damage to the other side of the brain. This is particularly true of the haematoma in the optic lobes or striate cortex and the adjacent regions, or in one of the two lobes alone: it is rarer in the case of injury to a site lying closer to the convolutions or the cerebrum, and even rarer at sites lying behind these lobes, towards the medulla oblongata.[15] The flow of blood elicited by tearing of a blood-vessel at first penetrates a large area of the brain, usually producing complete unconsciousness and other signs of a massive lesion of the central nervous system. The haemorrhage also causes after-effects more extensive than the trauma itself would have produced, e.g. inflammation of the brain. But when these effects have subsided, we are left with phenomena resulting from the blockage or destruction caused by the haemorrhage itself. The peculiar feature of these is that the ability to sense is weakened or even completely absent on that side on which the brain is *not* damaged: this applies to the upper and lower limbs, more rarely on only one of them, *but particularly to the trunk from the dorsal round to the ventral mid-lines.* [56] *Moreover, those muscles subserving movement in the paralysed limb or limbs are also more or less paralysed. But this does not apply to the muscles which move the trunk in opposite directions.*

Also very frequently seen in this connection are a paralysis of one half of the tongue and of several facial muscles, and a loss of tactile sensation of the same side of the head as that where the limbs are paralysed.

I have carried out a series of experiments on a few patients in order to determine in detail which muscles were paralysed and whether completely so: I have also investigated the extent and degree to which there was loss of tactile sensation and common sensibility. I have reached the following conclusions. In the simplest uncomplicated hemiplegias *only the touch organs of one side and their controlling muscles* (these only incompletely) are paralysed: muscles not concerned in movement of the touch-organs are free of paralysis. Thus the paralysed regions are in the skin and the muscles of one arm (including among others the cucullaris, latissimus dorsi, and serratus anticus major), the muscles of one leg, one half of the tongue, and one half of the lips, and the remaining areas of skin on the head and trunk of the same side. *On the other hand, those muscles not contributing to movement in limbs provided with a touch-sense, but which are concerned with moving one part of the trunk against the other, are not paralysed:* this includes the muscles mediating stretching, bending, and turning of the trunk, and the intercostal and abdominal muscles. It must be understood that the paralysis frequently does not extend over the whole of one side of the body, but is restricted only to the arm or leg or to a region of the skin of the trunk. But in uncomplicated cases, it does not extend to the muscles on the other side: the ventral and dorsal mid-lines form a limit beyond which the paralysis of the skin does not penetrate. [57] The arms, legs, tongue, lips and jaws are regions not formed immediately at the beginning of man's development, at the time the head and trunk commence, but grow later as appendages. The nerves of these touch-organs and their muscles appear to extend high into the cerebrum, whereas the nerves of the muscles moving the trunk appear to have their central point far below in the medulla oblongata, and therefore do not accompany the touch-nerves into the brain. As certain regions of the skin and the muscles named are simultaneously paralysed, and after the illness has ameliorated, are often found to be free of paralysis: and as, in the rare exceptions named, muscles can be paralysed without concurrent traces of loss of sensation in the skin and *vice versa*, so that these ostensible exceptions require detailed examination: we might justifiably assume that at the site of the haemorrhage in the brain, the nerves from one whole side of the skin lie close together. For if this were not the case we should not expect to find, with anything like the frequency we have, that such a localized lesion could lead to signs of paralysis in the skin of the whole side of the body.[16] Moreover we may assume that the nerves of single regions on the skin must lie close to the nerves of certain muscles, at the injured site of the brain: the pressure exerted by the haemorrhage, or the disturbances arising from the interpenetration of the blood, nearly always affect these specific nerves of the skin and of the muscles. I am of the opinion that (at the site of the hemiplegia-producing haemorrhage) we should not look for the site of the centre for touch, but suggest that the nerves at this point are still en route to this centre. [58] If the nerve-centre itself is affected convulsions seem to occur, associated with loss of sensation in the affected regions. In cases where I have examined patients afflicted with hemiplegia the skin and muscles were not always paralysed at all sites, nor is this to be expected. For the disturbance wrought by the pressure of the outflowing blood or by its penetration of the brain certainly does not affect every single nerve-fibre nor all the bundles of fibres in their vicinity. If we simply exert pressure on the ulnar nerve, we do not affect all the nerve-fibres, or all nerve fibres equally. In all these cases we say the feeling is 'fuzzy'*. Because many nerve-fibres blocked or restricted in conduction capacity are intermingled with others which are not so disrupted, the sense of touch seems to be inordinately disturbed. This can also be seen in hemiplegic patients with respect to their motor ability: however muscle-functioning appears to be even more disturbed thereby than does skin-functioning. On the skin, regions insensitive to pin-prick or to contact with a spoon which has been dipped in hot water alternate with regions in which equally strong stimuli may be felt as contacting but not as painful. But it would be a mistake to imagine that the ability to perceive pain had been eliminated from these regions, while the touch-sense persisted: pricking and touching with hot objects only fail to evoke pain because they are felt merely as a weak tingling, which could not be called painful. Hence they do not produce a sensation of touch so much as a sensation of 'common sensibility'; this however is too slight in strength to cause pain because only a few fibres transmit the sensation, or because the fibres transmit the sensations only incompletely and faintly. But on many regions of the skin the sense of touch remains undiminished, and at these sites the patient does feel pain upon pricking or touching with a hot object. The disturbance suffered by the touch-sense on account of the fact that paralysed and unparalysed nerve-fibres are mingled together may also be demonstrated by the fact that the patient cannot distinguish between a touch by a finger and a touch by a pin. [59] Moreover he has a very vague concept of the locality at which he has been touched and he believes, for instance, that he has been touched on the calf when it was actually on the instep, or on the small of the back when it was actually in the middle of the back. Whether or not the touch-sense is intact can be easily determined by finding out whether the patient can distinguish between moderately warm and moderately cold objects.[17]

Even the way in which the touch-nerves end in the skin is not yet fully known. [60] My experiments on localization in the skin make it very probable that each elementary nerve-fibre makes a large portion of the skin sensitive, but it is not yet certain by microscopic investigation how this occurs, whether such an elementary nerve-fibre bends frequently in all directions, recording stimulation at several places in its course, or whether

---

* German, *pelzig*.

it branches in the neighbourhood of its ending. Further investigation must still decide, e.g. whether the sinuosities observed in the twisting elementary fibres of the dental nerves, described and illustrated by G. Valentin[18] in the molars of the sheep, really are the endings of these nerves, and also whether the skin-nerves end in this manner in the skin-papillae, hair-bulbs, and other parts of the touch-organ. The observations of E. Burdach and Valentin[20] on the touch-nerves in the skin of the frog do not suffice to yield a concept of the effects of warmth, cold, and pressure on the touch-nerves. [61] Branching of the nerve-endings has occasionally been observed in Vater's or Pacini's corpuscles, but it is still doubtful whether these are sensory organs. Actually this branching has been observed in the Pacinian corpuscles in two cases, by Henle* and Kölliker,[21],** and so many times by Herbst[22],*** that Herbst is inclined to accept such as a regular phenomenon. These corpuscles, discovered by Vater[23],**** in 1747 in human hands and feet and called *papillae nervae*, recently renamed after Pacini, are very puzzling regions, and the fact that Henle and Kölliker have discovered their regular occurrence in the mesocolon and pancreas of cats, is inconsistent with any idea that they be considered as apparatus connected with the tactile sense. Herbst[24] found 20-160 Pacinian corpuscles in the mesocolon of cats, 2-79 in the mesentery, 40-50 on the surface of the mesenterial glands, and 40-60 in the pancreas. He found none in the leopard. Herbst calculates that there are 600 Pacinian corpuscles in man's hand: he counted 223 in the palm, 65 in the thumb, and 95 in the index finger. They lay in the panniculus adiposus. They have been sought in vain in the soles of the feet of rabbits and hares; Herbst could not find them in the soles of polecats or weasels, and remarks that a fibrous padding of fat is also lacking in this animal in places where they are found in animals and man. Higher up he has found them in cats, dogs, guinea pigs and even rodents, on the skin of the inner surface of the radius,

---

* F.G.J. Henle (1809-1885), Göttingen anatomist who gave his name to many kinds of anatomical tissue.

** R.A. Von Kölliker (1817-1905), Anatomist and histologist of Zürich and Würzburg. With Henle, he introduced the term 'Pacinian corpuscle" in 1844.

*** E.F.G. Herbst (1803-1893), Göttingen anatomist.

**** A.V. Vater (1684-1751), Professor of Anatomy at Wittenberg.

by the nervus interosseus. The ending of the nerve-fibre, connected to which is a button-like swelling, lies in an oval, transparent capsule of many concentric layers: Herbst considers these to be continuations of the layers composing the thick sheath of the elementary fibres of the touch-nerves. [62] These layers seem to be so loosely arranged about the oval capsule that they have fluid-filled spaces between them. In a young hedgehog Herbst counted 14 concentric layers about the oval capsule, and 20 in the guinea pig. Where the nerve-fibre divides, there are found just as many capsules, their layers being connected continuously with and merging into, each other.

The sensitive surface of thick skin and its countless swellings or papillae, as well as the organs of touch such as hair-bulbs or the dental pulp lying in or below the skin, are well supplied with blood-vessels and nerves and are protected by a bloodless and nerveless cover from over-intense, destructive, painful stimuli. This cover is sometimes thin, as in the epidermis of the conjuctiva of the eye, or of the tongue, and sometimes very thick as in the epidermis of the palm and the hollow of the feet. Sometimes it is so thick and prominent that, probe-like, it conducts impressions to the nervous regions through a long section of insensitive material, as in the case of teeth, hair, and nails. If these protective insensitive coverings are removed, the lightest touch or considerable propagation of heat or cold arouse pain, as in the exposed dental pulp, or skin stripped of the epidermis. The horny coverings conduct heat very slowly. Regions provided with a thin epidermis therefore convey sensations of heat and cold more rapidly and more intensely than do regions with a thicker epidermis.

The skin swellings, papillae, are small vascular protruberances on the thick skin, invisible to the naked eye, as their height is only about 1/25 Paris *line*. Each papilla is divided, my investigations have shown, into 2, 3, or several smaller swellings from which many others diverge, so that their points are fairly well separated from each other and stand out from the points of the neighbouring swellings. This holds both for the palm and for the back of the hand with the difference that the swellings in the palm are arranged in rows forming curving and easily visible lines of elevations, while the large swellings on the back of the hand do not form rows. [63] I found that the raised lines in the palm were an average of 0.23 Paris *lines* wide at the base. Such lines border closely on their neighbours, and each contains about two adjacent rows of swellings, each subdivided into 2, 3, 4, or more smaller swellings. This meant that one could count at least 81 large swellings or 150 or 200 small swellings in each square Paris *line*. These skin-swellings lie in the innermost, soft, opaque layer during this development of the epidermis; in white persons they are whiter, in dark-skinned persons darker, than the surrounding; and they form what has been called the Malpighian reticulum. This innermost layer of the epidermis does not develop an equally thick cover around the individual swellings, but only a fairly thick general cover over each elevated line of the palm, with as many hollows in its inner surface as there are swellings. If the inner surface of the Malpighian reticulum is observed when the epidermis including the Malpighian reticulum has been completely separated from the thick skin, then the various parts of the Malpighian reticulum, which now have their deepest depression between the swellings and surround the hollows containing the large swellings, form a network which appears white in white-skinned persons and dark in dark-skinned persons. This net-like formation in the inner surface of this layer probably led Malpighi to call the whole layer a net, which actually is not the case. The Malpighian net may be distinguished from the fully developed layers of the epidermis by the fact that the latter consists of countless, extremely thin lamellae bending about in parallel, each of which is made up of flattened elementary cells growing into each other at their edges. In the Malpighian reticulum the elementary cells are not flattened to the same degree, in fact the innermost are still rounded, nor have the elementary cells grown into each other at the edges in such a way as to form definite lamellae consisting of a single layer of cells. I found, for the palm, that the thickness of the Malpighian net together with the fully-formed epidermis was about 1/4 Paris *line*; the maximum thickness of the Malpighian net was 1/22 Paris *line*, and its minimum, above the parts of the swellings, was 1/44 Paris *line*. [64]

The superficial layer of the thick skin is much richer in capillaries than its remainder, apart from the pores and the hair-bulbs. It is traversed by a thick net of blood-supplying capillaries whose cross-sections I have measured both in the skin of the arm of a man, when they were completely filled with blood, and in the instep of a child whose capillaries had been completely injected. The blood-filled capillaries had the following diameters: the thinnest was 0.0056 Paris *line* i.e. about 1/178 Paris *line*; the thicker ones were 0.039[25] or 1/77 Paris *line*. The spaces surrounded by the capillaries in the net were larger, at some places their diameter was twice as large as the diameter of the capillaries, at others only 1/3 larger.

Out of this dense network of capillaries a capillary extended to each small swelling, which twisted back on itself near the point of the swelling, and returned to the network, forming a simple capillary loop within each swelling. Apart from this loop the swelling had no other blood vessels. Sometimes the capillary loop was smooth, sometimes it snaked about.

It only remains to describe the nerves in the swellings in the same degree of detail. I hope that in the future I shall succeed in making these visible: so far, however, I have not yet managed to do so.

I. Localization in the skin

The sense of localization consists in our being able to distinguish between two sensations, which even if equal in all other respects, are excited at different places on our body or in our organs of touch. If pressure or the effects of heat or cold evoke a sensation, we can approximately designate the site on the skin at which the stimulation evoked the sensation; and if we simultaneously or successively receive impressions of warmth, cold or pressure from two sites on the skin, we can, provided they are not too close to each other, distinguish the two sites at which the stimulation occurred. [65] We can also indicate the larger or smaller distance between the two sites and give the approximate direction of an imaginary line joining them. 20 years ago[26] I reported in a series of experiments just how accurate this ability is and discovered that it is present in very varying degrees in different parts of the skin; for example, it is 50 times more complete on the tip of the tongue than on the skin covering the middle portion of the upper arm or ankle. The method of investigation I applied was as follows: I took two small objects of equal shape and touched two parts of the skin on persons whose eyes were closed or averted. I asked them if they felt whether one or more objects were touching them, and in what direction the line ran, which they imagined to connect the contacted parts of the skin, i.e. whether it can lengthwise along the body, or across the body-axis. For this purpose I ground down the points of a cylindrically-legged pair of compasses until the ends had a diameter of 1/3 Paris *line*. When these touched the skin, they would then not pierce it, but would give rise to a pronounced impression of touch. The reason for this was that if a contact evokes pain, the observation thereof becomes less defined: for pain is never so well localized as a painless but sufficiently strong touch with a not too small surface and which does not cause pain. [66] At first the compass-legs were set well apart: but by gradually reducing their distance, I succeeded in finding that setting of the two ends of the legs where their two impressions begin to feel like a single impression. Even then the observer could often determine whether the line joining the ends of the compass points ran in the direction of his body-length and limbs or across it. For although only one impression was obtained, the touched part of the skin appeared to take on a length-wise configuration, and the subject could say which were the long and short axes of this long configuration of his skin. By recording the distance between the compass-points at which two touches could be distinguished, or at which at least the direction of the compass legs could be determined, and by repeating the experiment at other times, gradually extending the investigation to various parts of the skin, I could obtain a general picture of the accuracy of the sense of touch, insofar as it could be judged by the ability to localize. It was of the greatest importance to avoid fatiguing the subject, and therefore not to continue too long with the same experiments.

Among the results were the following. If the compasses were opened to 3/4 inch, and they touched the skin at the posterior part of the zygomatic bone in a transverse direction, only one contact was felt, or at most the subject perceived that the ends of the compass were very close together. But the closer we came in these experiments of touching the middle of the upper lip, the farther apart the compass-points seemed to be, and the more clearly the double contact was felt. The circle-points therefore seemed to be furthest apart when the middle of the upper lip intervened between them. Similar results were obtained when the ends of the compass were placed in a vertical direction and finally came to touch the middle of the upper and lower lips simultaneously. If the experiment were changed so that the two ends of the compass were arranged vertically to touch the cheek just in front of the earlobe: then the two ends were moved simultaneously and transversely across the face so that the two ends came to touch the middle upper and lower lips, and the movement was continued round the other half of the face towards the opposite earlobe: the observer did not have the impression that the two ends of the compass described two parallel lines, but that, the closer they approached the lips, the further apart they became, narrowing again as the movement continued round the face, until finally only one point of contact was felt. [67] The upper end of the compasses appeared therefore to describe an arc bowed convexly upwards, the lower end an arc bowed convexly downwards. The same experiment can very easily be carried out on the hand also. If the compass points are set at 4 or 6 Paris *lines* apart and touch the middle of the lower arm transversely, and the compasses are then moved in a steady and even contact with the skin towards the palm and on up to the end of the index-finger, the compasses appear at first to describe a single line: but on the hand this divides into two lines, and the closer they get to the end of the index-finger, the more the compasses seem to be opening and the more divergent the lines they describe apparently become. The same can also be experienced on the tongue if the compasses are set at two Paris *lines* and then steadily brought from a transverse position in the middle of the tongue down towards the tip.

My explanation of this remarkable phenomenon, which has already been put forth in my earlier report,[27] is as follows. [68] *Fontana* has observed that the elementary fibres of the nerves are single fibres neither giving off nor receiving branches. On the basis of this and the experiments I myself have done on the variability in accuracy of the sense of localization on various regions of the skin, I assume that *when two otherwise equal impressions simultaneously excite the same nerve-fibre at various places, only one sensation, not two, arises.* In fact each elementary nerve-fibre must sensitize an area of skin much larger than the area of its own cross-section: for if one imagines all the sensory nerves cut straight across and piled on top of each other like cords of wood, the total cross-section of all these nerves would be very much less than the surface area of all the skin. We have already experienced considerable difficulty in explaining how it is that the skin could be so sensitive at all its points as to make a prick with the finest English sewing-needle perceptible everywhere. *Prochasca** has suggested that the nerves widen at their ends in such a way that the nervous tissue penetrates the skin-substance; and *Reil*** assumed that the ends of the nerves were surrounded by a nervous 'atmosphere' similar in kind to the electrical atmosphere surrounding an isolated conductor charged with electricity. In our present state of knowlege on the admittedly little investigated question of the mode of endings of the nerves, we are obliged to assume, as mentioned earlier, either that the ends of the elementary nerve-fibres sensitize a large portion of skin *by winding and twisting about so that one and the same fibre is receptive to touch at several points along its sinuous course*, or that this view of Fontana's on the peripheral nerve-endings is invalid, and that the *elementary nerve-fibres divide into branches near their termination*, thereby innervating many points on the surface of the cutis vera. A final alternative is that both types of arrangement could act together to ensure that a single elementary nerve-fibre sensitizes a whole region of skin. [69] In favour of the view that elementary nerve-fibres twist about are the findings of

---

* G.P. Prochasca (1749-1820), Professor of Anatomy at Prague and, later, Vienna. He is best known for his research on reflexes and on the vascular system.

** J.C.R. Reil (1759-1815), Professor of Medicine at Halle. He is best known for his work on brain structures and as being a pioneer of German psychiatry.

Valentin, mentioned above; in favour of the view that they branch is the fact that Henle, Kölliker, and more recently Herbst in frequent instances, have observed that the elementary nerve-fibres often do divide into branches in the Pacinian corpuscles: and R. Wagner and others have noted that such an arborisation of nerve-endings can be seen in the electrical organs of fish. R. Wagner has also observed that the elementary threads of motor nerves divide in the vicinity of their endings.

But no matter how the elementary nerves do extend to cover the skin, the suggestion may be put forward that the skin is divided into small *sensory circles*, i.e. into small subdivisions each of which owes its sensitivity to a single elementary nerve-fibre. Now my investigations have shown that two stimulations of similar kind applied to separate sites within a single sensory circle on the skin are felt as if they were made at one and the same site; and moreover, that the sensory circles of the skin are smaller in regions provided with an accurate touch-sense and larger in areas provided with a less accurate touch-sense. In order for two impressions made simultaneously on the skin to be localized as two impressions lying a certain distance apart, it appears necessary that the impressions are not merely made upon two separate sensory circles, but also that between these sensory circles there be one or several circles upon which no impression is made.

The actual shape of each sensory circle has not as yet been completely determined. But it is highly probable that the sensory circles on the arms and on the legs have a longish shape and lie in a position such that the long axes lies along the lengths of the limbs; for to touch the limb with the compass points lying in a direction parallel to the long axis of the limb does not give the same results as if the imaginary line joining the compass points lay at right angles to the long axis of the limb.[70] In the former case the compasses must be opened to a much larger extent than in the latter, if the contacts of the two compass-legs are to be felt as two sensations. For example, in the middle of the upper arm, both anteriorly and posteriorly, a vertically held paid of compasses must be opened to almost twice the extent of that required for a horizontally held pair, in order to elicit two sensations of contact clearly. But even on the upper leg, lower arm, and lower leg, the sense of localization was much finer in the horizontal as opposed to the vertical plane. In many other parts of the body this difference is not found, which would lead us to assume that in these regions the sensory circles have a shape which is approximately round.

If my explanation is correct, these equally large sections of the skin at those places endowed with a very accurate localization-sense must possess more elementary nerve-fibres than those regions with a very coarse localization-sense. This also can be confirmed: if we take the two thick nerves on the volar side of the hand and the two thin ones which run along the back of each finger, and then compare the nerves of the fingers with the few nerves innervating an equally large area of the skin on the back, we find that the number of elementary fibres in the sensory nerves on the volar side is much larger than that on the back side of the fingers, and is least of all on the back itself.

By long usage and repeated motor activity of our limbs, with their touch-sense, we have come to form a dim awareness of the number and position of our sensory circles. The more sensory circles lying between the two points of a compass touching us, the further apart the points seem to be and *vice versa*. If therefore the sensory circles are small and numerous, as on the finger-tips or the palm of the hand, or equivalently, very many elementary nerve-fibres all terminate in the same square inch of the skin-surface, then the compass-points touching these areas seem to be further apart than if they touched part of the dorsal skin, in whose surface only a few elementary fibres are found to terminate within a square inch. *The number of elementary nerve-fibres ending within a square inch of a touch-organ endowed with a localization-sense influences the criteria by which we estimate filled space.* [71] If, as is the case for infusoria, we were a hundred thousand times smaller but remained proportionally organized just as we are now, with our skin and nervous tissue arranged into just as many divisions within which two stimulations could be discriminated as two spatially separated impressions despite their smaller scale: and if our movements remained proportional in amplitude and speed to the reduced size of the limbs, despite being less extensive and slower; then if the surrounding world of objects was also proportionally smaller and denser, and moved proportionally more slowly there would be no difference between the life we now lead and that which we would lead under those circumstances. But we would find ourselves much less suited to dealing with relationships between bodies as they now exist. A drop of water an inch long would be about 1700 times longer than ourselves and appear to us like a large pond. Our sensory view-point of the world depends on the smallest standards we have for measuring time and space. The smallest standard we have for measuring time, is the smallest interval within which we can perceive a change, and is discovered when we know how many times we can use our will within a second, or how many distinctive sensations we can have within a second. The smallest standard we have for measuring space depends on how many distinguishable sensations we have on a square unit of surface. For example, in vision, we use the smallest discriminable black and white squares; in touch, we measure the smallest possible square elevations or depressions discriminable. This last ability to discriminate many parts of a square depends, as I have argued, partly on the number of elementary fibres of the nerves ending within a square of skin; but it also depends, in fact to a greater degree, on the number of nerve-fibres ending in the retina of the eye; for in the central portion of the retina the ends of the elementary fibres are much more densely packed than in the skin, and we therefore, by way of our eyes, are able to perceive many more discriminable areas within a square Paris line than we can by the skin. [72] So we utilize not so much the measure provided by the touch-sense as that provided by the visual sense; and we attempt to reduce sensations obtained from the touch-sense to the same standards as those obtained *via* the visual sense. A person born blind however can only rely on the touch-sense. Despite the fact that he can, by mental elaboration, reconstruct geometrical relationships, the space of an inch must be filled by many fewer discriminable parts, and therefore seem smaller than it would to a sighted person; for the blind person has no means of extending his perception of the space contained by an inch beyond that conveyed by his touch-sense.

Volkmann[28]*, has expressed his essential agreement with me. He says "if the ends of a compass are set about an inch apart, and first placed on a finger-tip but then gradually moved in and up towards the centre of the body, the distance between the compass-points appears to diminish: and there is a position on the skin at which this distance is felt to be longer than the distance of a Paris line at the tip of the finger. This point is the one at which the distance of an inch is the smallest distance which can be sensorially perceived. Thus the skin judges the size of objects by taking the size of this last perceptible distance as a unit of measurement. If we call this unit of measurement x, then the size of an inch at the finger-tip is 12x, and that at a site in the middle of regions of the arm is 1x; for each point on the skin designates to a touched object a size corresponding to the number of sites included at which an object of size x can be discriminated in isolation."

J. Müller[29] is also inclined to agree with my viewpoint, but objects that according to this view, the two sensations arising if the finger-tip touches the arm should not be the same; the finger-tip should feel the area of contact with the arm to be large, and the arm should feel the area of contact with the finger-tip to be small. [73] But these two sensations are certainly not equal; one is clear, and the other is faint. In this case the clear sensation is relied on, and the attention cannot be switched from it to the faint sensation in a manner

* A.W.V. Volkmann (1800-1877), Professor of Physiology at Halle. He is best known for his studies on blood flow, the sympathetic system, and vision.

allowing the judgement on the extent of the touched surface to be based entirely on the faint sensation. Müller is of the opinion that my observations could also be explained if the sensations of feeling in the arm were so blurred as to form dispersed circles*, those on the finger-tip being, in contrast, sharply defined. My experiments cannot be explained in this way. If the two points of the compasses are drawn over the lips so that the red parts of the lips and the mouth orifice lie between them, the perception of distance in the region of the corners of the mouth is quite clear, yet the distance between the compass-points seems to increase greatly as they moved towards the centre of the mouth. The 'dispersed circles' in the eye give us the impression that the object is greater, but on the arm a whole depressed area of skin seems to be smaller than on the hand. 'Dispersed circles' cannot reduce the apparent distance between two compass-points, but only make them less distinct.

My remarks on the touch-sense may also be applied to the eye. In the tongue, the sense of localization is finest at the tip, decreasing as we move further away around the tongue: in the eye it is similarly most sensitive where the retina is sectioned by the axis of the eye, gradually decreasing as we move peripherally in any direction from that point.

If one takes the end of a sectioned cylindrical, or four- or three-sided prismatic brass rod, 1, 2, or 3 inches in diameter, and presses it against a region of skin poor in localization, e.g. the middle of the upper or lower arm, the subject, as will be shown further below, does not perceive the shape of the object pressing on him: he feels he has been pressed by a solid object of undefined shape. But he can easily tell the shape by using his hand or tongue. Using the tip of the tongue, he can even perceive the shape of the cross-section if the brass rod has a much smaller diameter, e.g. $1\frac{1}{2}$ Paris *lines*. [74] It has been long known that objects forming retinal images from the side of the optical axis are so dimly perceived that the outstretched fingers of a hand cannot even be distinguished or counted if presented from the side. Hueck,[30],** who has attempted the admittedly tentative measurement of the fall-off of sensory acuity of the retina from its midpoint out to the periphery on the various sides, has found that the diameters of the smallest image on the retina which can be seen are:

---

* German, *Zerstreuungskreise*.

** Probably A.F. Hueck (1802-1842), Professor of Anatomy at Dorpat who specialized in the study of vision.

at the centre	0.0008	Paris *line*
5 degrees from the centre	0.0024	" "
14 " " " "	0.0060	" "
25 " " " "	0.0030	" "
50 " " " "	0.0340	" "

For an object only 5 degrees from the centre to be seen, it has to be three times as large in diameter as an object visible centrally.

Volkmann has demonstrated that the cause of this is not to be found in any incompleteness of the image as presented peripherally, and I must confirm this. The exposed eyeball of a recently killed white rabbit shows that the image penetrating through it from the side is quite clear and in no way distorted or blurred. I suggest that the reason for the poor sensitivity shown is that the endings of the elementary fibres of the optic nerves are exceptionally densely packed at the area of the so-called central depression, and that the further away the fibres are from the midpoint of the retina, the more scattered they become. This assumption would explain how it is that the elementary nerve-fibres contained in the optic nerve can sensitize the whole retina yet give exceptionally fine localization to the middle regions thereof. According to Volkmann's estimates, the cross-section of the optic nerve is about 50 times smaller than the surface of the retina, so that each nerve-fibre must extend over an area of this surface in the middle which is 50 times greater than its own cross-section. This could happen in the same manner as in the skin, each elementary nerve-fibre belonging to a peripheral section of the retina spreading out to form a sensory circle, either by sub-division or some other means. [75] These sensory circles would be largest at those points in the retina most removed from the centre. I find that the concept given by physiological investigations of structure of the so-called central depression and the rest of the retina is highly consistent with that provided by Grube's microscopical observations. Grube[31] studied the eye of a man who had died but a few hours previously from rupture of the spleen. At this time the retina still adhered firmly to the vitreous humour, while later it could easily be detached. The yellow portion of the retina, along with the vitreous humour, was put under the microscope, being not compressed, but merely covered with an extremely thin glass-plate 1/4 sq. in. in area. The yellow spot, under 300x magnification, looked like the shagreen used for covering small boxes. The longish cylindrical bodies which compose the retina at this point were arranged in a very orderly manner and became smaller as they approached the centre. In the middle they were only 1/4 or 1/5 the size of the bodies found outside the yellow spot. At the point of transition from the yellow spot into the retina these bodies spread themselves out like the rays of a star, the distances between each other varying irregularly. At this point they were not only larger but also less clearly defined in outline. Unfortunately Grube did not make any microscopic measurements. In order to estimate how far the number of elementary nerve-fibres contained in the optic nerve could spread out so as to occupy the most sensitive part of the retina with thickly packed nerve-endings, we must guess that each sensitive region could have a diameter of only 1/2-1/3 Paris *line*. My brother, Wilhelm Weber, and I have determined the size of that sensitive part of the retina, most highly endowed with a sense of localization, in the following way: I find that if I look straight at the writing of this publication[32]*, and fixate one letter on one line, I cannot simultaneously recognise the letters lying above or below that line, in the lines above and below it; however if I gaze directly at the mid-point between two lines, I can recognize the two single letters of the two lines bordering the space I am fixating. [76] Within a single line I can recognize about 3, or at the most 4, neighbouring letters at once, keeping my eye stationary. These experiments have the disadvantage that the eye moves a little almost voluntarily in order to see more than is possible with the eye fixated in one position. To avoid the uncertainties arising from this we read larger script by the light of an electric spark, and to avoid the possibility of guessing, we used a language whose words were not known to us. As the light arising from the discharge of a Leyden jar is momentary, so that there is no time to move the eye, but only to perceive that which forms on the most sensitive part of the retina, the question is raised whether there actually is sufficient time in which to perceive the momentarily illuminated script. But there is, for a produced light impression can persist, according to Plateau,** for 0.32-0.35 seconds, and according to Hueck, for 1/6 seconds. The diameter of the most sensitive part of the retina could then be measured, knowing the size of the surface on which the letters could be recognised, the distance of these from the eye, and the distance of the focus of the light-rays from the retina. This focus was taken to be just in front of the midpoint of the eye. For myself and my brother, this diameter came to between 1/2 and 1/3 Paris *line*. If it is accepted that only a part of the retina, about 1/3-1/2 Paris *line* in diameter, is thickly populated with thin nerve-

---

* In the original edition the space between successive lines of type is about $2\frac{1}{2}$.

** J.A.F. Plateau (1801-1883), Professor of Physics and Anatomy at Ghent who studied visual sensation and, later, hydrodynamics.

endings, and that the nerve-endings are particularly thick within this region at its centre, the central depression, then enough nerve-fibres will remain over to cover all the rest of the retina with sensory circles, these increasing in size towards the periphery. [77]

Joh. Müller[33], also, assumes that a single elementary nerve-fibre cannot serve to distinguish between two impressions. The objection he himself raises to this suggestion, namely that a pressure produced upon the trunk of the ulnar nerve is felt not only at the site of the endings of the pressed nerves but also at the point of pressure on the trunk, does not seem to argue against this assumption. The pressure exerted at the region of the elbow is clearly felt by way of touch-nerves (branches of the cutaneus internus) ending in that region; and since nerve-pains are never exactly localised at the site of their origin, nor clearly delineated in area, the pains in our present case are ascribed also to the region at which the touch-nerves are excited by the painful stimulus. There are many other instances where more indefinite sensations are wrongly localised at sites at which we simultaneously perceive sensations from the touch-nerves. We imagine, for instance, that we hear the tone of a tuning-fork via a tooth to which the handle of the tuning-fork is applied. But the determination of the source of a sound is so incomplete in the hearing-organ that such a fine judgement cannot be made therewith. As I showed 20 years ago for the first time, if one ear is blocked by the finger, the tone previously perceived in the tooth, now seems to sound in the blocked ear, probably because under these circumstances the highly sensitive tympanum of the blocked ear is violently vibrated.[34] On the other hand Volkmann[35] has expressed himself as opposed to this assumption. He believes he has been able to prove that even if one and the same nerve-vessel in the retina is stimulated by light at two sites, the two sensations may be discriminated from each other in terms of their sites of excitation. He asserts that two impressions may be distinguished even if they lie only 1/1000 inch apart on the retina. [78] But the retinal elements are supposed to have a larger cross-section, and have therefore been equipped to produce at least two discriminable impressions. According to Smith, the smallest visual angle at which many persons can distinguish two points is about 40 seconds, and from this he calculates that the smallest sensitive point on the retina must have a cross-section of about 1/8000 inch or 1/666 Paris *line*. According to Hueck's[36] measurements a white lustreless point on a black field will disappear if the visual angle it subtends is only 10 seconds. But Hueck himself correctly rejects this method, for one can even see fixed stars of no measurable diameter, and also others of such dimness that they do not shine. When Hueck observed a black point on a white background, it disappeared from sight when the visual angle at which it was seen was only 2 seconds. But even this kind of experiment does not reveal what is desired: no conclusions as to the diameter of the smallest retinal nerve-fibre can be arrived at from these experiments. For why should a black spot which is so small that its image does not even completely cover the end of a nerve-fibre, not also be visible, if the impression is strong enough? If the nerve-fibre in question is simultaneously stimulated with white light, the point will appear somewhat paler, but it is quite possible that it is still perceptible in contrast with the brighter illumination of neighbouring nerve-fibres. On the skin light contact with a pin point can be felt anywhere, but two simultaneous contacts 1 or 2 inches apart on the back are only perceived as a single contact. Perhaps something similar is also observed on the retina. The method applied by Smith must therefore be adhered to. When Hueck applied this method, observing two black points on a white background, he found that they fused to a single point when the visual angle subtended (angle of the space alone, or of the space plus the two points?) was 64 seconds. [79] According to this the smallest visual angle at which two points could be distinguished was more than 1/3 larger than that of Smith, according to whom this angle was 40 seconds. Volkmann has taken, not two points, but two parallel lines, and was able to distinguish them when their images on the retina were only 0.00014" (i.e. about 1/7142 inch, or 1/559 Paris *line*) apart. Valentin[37] was able to distinguish two lines whose retinal images were only about 0.00009" (i.e. 1/11000 inch, or about 1/1000 Paris *line*) apart. But I do not consider it irrelevant as to whether points or lines are being fixated, nor can I agree with Volkmann's conclusion that the ends of the fibres of the visual nerves are too coarse to enable such a discrimination to be made. Volkmann says: "If two threads of a spider's web are stretched close together over a little frame, and removed to a distance at which the duplicity of the threads is still just distinguishable, they are still distinguishable as two threads no matter in which direction the frame is turned or pushed. But granted that, for a given position of the eye, the images of the two parallel lines could have fallen, on definitely different fibres, this would not be the case for every possible orientation of the lines: for it is impossible to conceive of an arrangement of nerve-endings such that straight parallel lines set at a distance less than the diameter of these endings, could stimulate different endings no matter what their orientation. Rather, the images of such lines must infallibly retraverse one and the same nerve-fibre as the lines are moved back and forth, giving a sensation at such a point of one, rather than two, lines. We would thus perceive a figure in which the two lines would alternately fuse into each other and separate again. As this is not the case, the same nerve-fibre must be capable of serving for the apperception of two discrete sensations. According to this view we might even suspect that a single elementary fibre is capable of inducing the perception, not merely of two, but even ten or more different points." [80] As I have said I cannot share this opinion. I have found, in the trunk of the human optic nerve 18 hours after death, many non-varicose elementary fibres whose cross sections varied between 0.0007 and 0.0001 Paris *line* = 1/1428 - 1/1000 Paris *line* (in inches, 1/17000 - 1/12000 Paris inch). Admittedly I have not measured the endings of these fibres nor does anyone know anything certain about them: and naturally they must not be measured in the peripheral sections of the retina, but in the so-called central depression. But so much is certain, that the picture I have given concerning the localization-sense in the retina is free of factual objection.[38] I must therefore persist in my earlier explanation of the value to be attributed to the behaviour of the elementary nerve-fibres, first revealed by Fontana.

It is known that the touch-nerves of our body spread out in such a way that they do not cross the middle plane, where the body is divided into its right and left halves. This interesting phenomenon is confirmed not only by anatomical evidence, but also by physiological observation, notably in hemiplegia: for the paralysis, in uncomplicated cases, does not cross the boundary between the right and left halves of the body: and as the nerves from two closely adjacent regions of the midline of the skin take quite separate and widely divergent courses, we would assume that touching two such regions would always produce two quite distinguishable sensations. [81] But this is not the case. This result seems to us to prove that local differentiation of sensations depends only upon the site of the nerve-endings in the skin or in the brain, and not upon the site at which the trunks of the nerve-fibres of the two sides diverge and at which the peripheral and central endings thereof are brought together.

We might imagine that the fineness and sharpness of the localization-sense, so marked in some skin-regions as compared with others, will partly depend on whether or not we are provided, via the eye, with exact knowledge of the distance of the various skin-regions from one another. Experience tells us, however, that this is not the case. If it were, those skin regions we never see would be provided by a much less acute sense of localization than those which we can visually observe. But this is definitely not so.

ACUITY OF THE LOCALIZATION-SENSE IN THE HEAD

In the head, the region supplied with the finest touch-sense is the tip of the tongue. Next comes that region of the lips forming the boundary between the red and the pale parts of their surface: the touch-sense here is almost finer than that of the finger-tips. Then follows the tip of the nose, then the eye-lids, then the upper edge of the eye-socket near the glabella, and the glabella itself. The most insensitive part of the face is that skin-area, just in front of the ear, where the jaw-bone branches off. Many parts of the skin of the head have therefore a highly sensitive touch-sense, and there are no regions which have the touch-sense blunted as is the case for many parts of the skin on the back, or some parts of the skin on the arms and legs. The head must therefore be declared to be that part of the body most outstandingly endowed with a fine sense of touch. [82]

On the head itself regions provided with hair have, not a finer, but a duller sense of touch than have regions without hair, e.g. the forehead. The hair-bulbs of the hairs of the head, so numerous and well-supplied with nerves as they are, are therefore not to be taken for touch organs affected by pressure, but are rather, as we shall show later, touch-organs which respond accurately to pulling. It may however be otherwise with the touch-hairs of animals, which are much stiffer, movable at will, and rooted in specialised capsules.

Let us go into more detail with this survey.[39] It is of great benefit to us that we have such a fine touch-sense in the mouth-orifice, on the tongue, and also that the teeth, which serve to crush food, are endowed with a touch-sense. Those parts incapable of crushing stand out thereby and as yet unchewed food can therefore be brought between the crushing surfaces of the teeth. The finest sense of localization of all, which renders the tip of the tongue superior to all other regions of the body, is limited to a very small area possessing a diameter of only 2 - 3 Paris *lines*. The under-surface of the tongue has a much less well developed sense of touch than does the upper surface. On the back of the tongue, the acuity of the touch-sense decreases, the further away the investigated area is from the tip. The anterior part of the gum has a very blunt sense of feeling: it becomes better developed however in a posterior direction in the upper jaw and on the hard palate. On the soft palate the touch-sense is even more sensitive than it is on the hard palate. The teeth act as probes and have a fairly fine touch-sense, but fairly hard pressure is required: this apparently operates on the sensitive skin carpeting the alveolus. [83] Its touch-sense is far superior to that of the gum; the mucous membrane of the cheeks is much less sensitive to touch than is the outer skin thereof. The lips are also much less sensitive to touch on their inner surface than on their outer surface. The touch-sense is most acute at the point where the red part of the lips changes to the part which is not red. Beyond this point the fineness of the touch-sense decreases and is least acute near the gums. The touch-sense of the outer surface of the upper and lower lips is more sensitive near the midline. After the lips, the tip of the nose has the most accurate localization-sense, finer than that of the lip or the bridge or the linings of the nose; then after the tip of the nose come the eye-lids, which are more sensitive to touch near the outer corner of the eye than in the inner corner. In this case, at the outer corner, the high sensitivity extends even to that part of the skin covering the junction of the zygomatic bone and the bones of the forehead. The region of skin on the upper edge of the eye-socket covering the corrugator supercilii is also still sensitive. Then comes the so-called glabella, i.e. the point of transition between the top of the nose and the forehead, and the forehead itself: for on the forehead the accuracy of localization decreases as we move upwards and rounds to the temples. Part of the cheek, well removed from the lips, has almost equal sensitivity. The part of the face in front of the ear, and the skin covering the branch of the lower jaw is less sensitive; but the middle of the chin and the nearby region of skin under the chin is more sensitive than the skin of the forehead, this sensitivity decreasing markedly however towards the neck, and being very much less pronounced on the neck itself. The outer ear, insofar as the localization sense is concerned, is one of the more insensitive regions of the head; but the skin inside the nose is even less sensitive to touch.

LOCALIZATION IN THE ARMS AND LEGS

The sharpest sense of localization in these areas is on the volar side of the hands, which markedly exceeds that of the back of the hands and is not inconsiderably greater than that of the plantar side of the feet. [84] The finger-tips, or rather the terminal sections of the fingers, on the inside of the hand, possess the most acute localization-sense, almost equalling that of the lips. This acuity decreases in the second section of the fingers, and more noticeably so in the first: but on the latter, the end directed towards the metacarpals is more sensitive than it is on the end directed towards the second section. On the metacarpal bones of the thumb, sensitivity is a little greater than on the metacarpal bone of the little finger. The sensitivity on the volar side of the second section of a finger stands in a ratio of about 5:2 to that on the back of the finger, and on the metacarpal bone the two sides stand in a ratio of

14:5. The least sensitive part is the skin round about the middle of the upper arm. If the compasses are set about 16-18 Paris *lines* apart, and are then used to touch the skin on the shoulder, acrominion, or elbow, i.e. over the olecranon, it is possible to tell whether the compass points lie longitudinally or transversely over the skin, and two contacts are felt. But on most of the upper arm, and on a small part of the lower arms, only a single contact is perceived and it is not possible to judge the positions of the compass-points: they have to be opened to a width of 2½-3 in. before this can be estimated. The skin of the upper arm is therefore somewhat less sensitive than the skin on the lower arm, and this in turn is far less sensitive than the skin of the hand. But the sensitivity does not decrease evenly between the hand and the shoulder: at the wrist and the elbow-joint it is somewhat greater than in areas intermediate between these points.

Much the same holds for the legs, except that the feet and toes are less sensitive on the sole than the hand and fingers are on their volar surfaces. Otherwise it is as has already been discussed above: on most parts of the arms and legs the ends of the compasses are felt more clearly and need be less widely opened if they are laid transversely across the skin as opposed to longitudinally.

LOCALIZATION IN THE SKIN OF THE TRUNK

The localization-sense is least well developed on the trunk. As we said earlier, there is not a single region in this area which has such a fine localization-sense as that of any part of the skin of the head, hand, or foot. [85] Even at the nipple, extremely sensitive in other respects and capable of erection following continued stimulation, the touch-sense is very blunted, a clear example of how the ability to experience a lively common sensation should be distinguished from the acuity of the touch-sense.

Localization in the skin of the trunk is most accurate at the two extremities of the back, on the upper part of the neck, and on the anus; the acuity decreases as we move towards the middle of the back. Localization is rather less accurate in the lower part of the neck than it is in the upper regions of the chest, and it is better developed round the navel and the pubic bones than in the regions between these.

On the side of the trunk regions at which the contacts of the compass points are felt more clearly when applied longitudinally alternate with regions at which they are more clearly felt when applied transversely. It is not yet certain however that experiments made on corresponding parts of the body in different persons will give the same result.[40] [86]

In the following table* I have shown, both in terms of Paris lines and of lines drawn to scale, the distances between the compass-points necessary for me to feel two contacts from the compass points as they touched me, or at least lead to my being able to judge whether the compass points were applied along the long axis or my body or transversely thereto.

	Paris Lines
Tip of the tongue	1/2
Volar side of the last section of the finger	1
Red part of the lips	2
Volar side of the second section of the finger	2
Dorsal side of the third section of the finger	3
Tip of the nose	3
Volar side of the capitulla ossium metacarpi	3
Midline of the back of the tongue, 1" from tip	4
Edge of the tongue, 1" from tip	4
Parts of the lips (not red)	4
Metacarpus of the thumb	4
Plantar side of the last section of the great toe	5
Dorsal side of the second section of the finger	5
Cheeks	5
Outer surface of the eyelid	5
Middle of the hard palate	6
Skin of the anterior part of the zygomatic bone	7
Plantar side of the middle foot-bone of the great toe	7
Dorsal side of the first section of the finger	7
Dorsal side of the capitula ossium metacarpi	8
Inner surface of the lips near the gum	9
Skin over the posterior part of the zygomatic bone	10

* This table corresponds to the one on p. 36.

Lower part of forehead	10
Hindmost part of heel	10
Lower part of the back of the head (hairy)	12
Back of the hand	14
Neck under the jaw-bone [87]	15
Crown of the head	15
Knee-cap and surrounding region	16
Os sacrum	18
Glutaeus	18
Upper and lower parts of lower arm	18
Upper and lower parts of lower leg	18
Back of the foot near the toes	18
Breast bone	20
Spine at neck below back of head	24
Spine in region of the 5 upper thoracic vertebrae	24
Spine near loins and lower thorax	24
Spine in the middle of the neck	30
Spine, middle of back	30
Middle of upper arm and upper leg	30

PERCEPTION WITHOUT LIMB-MOVEMENTS OF THE SHAPE OF AN OBJECT TOUCHING US [88]

We can tell the shape of an object or the distance between two objects in two ways: firstly, without moving our limbs, and secondly, by moving them. If the end, and therefore the cross-section of a cylindrical or three- or four-sided prismatic metal tube is pressed against the skin, without our being able to see it, an impression thereof is given on the skin and we can perceive its shape provided the cross-section is large enough. If cylindrical tubes of various diameters are applied, it is found that the end of the tube is felt as a solid object of indefinite shape provided it does not have a diameter exceeding the distance between the two ends of a sensory circle, the latter being defined as that distance, shown for various body regions in the above table, at which two impressions can be distinguished. Using a tube of only $1\frac{1}{2}$ Paris *lines* in diameter, I could feel the shape of a circle and its enclosed space only with the tip of the tongue, not with the lip or finger-tip; if the tube was 2 Paris *lines* in diameter, I could feel its shape with the middle part of the lip, and less clearly with the finger-tips, but not with the jointed parts of the last or second sections of the fingers. In order for the shape to be perceptible with these regions a diameter of 4 Paris *lines* was necessary: and at the first section a diameter of 5 Paris *lines* was necessary in order to feel it. On the metacarpus of the thumb 6-6½ Paris *lines* were required: on the abdomen the tube had to be 3¾ in. in diameter in order for it to be perceived as a shape. An even larger diameter was required for the back. We are only capable of discriminating round from three or four cornered shapes, without using touch-organs, by virtue of knowing the position of the skin-region being touched. We certainly do not have this knowledge innately, but arrive at it because objects more along the length of our skin and thereby excite successive sensations in certain rows of tactile points. It is most important, in order both to learn the positions of sensible points in our skin, and to be led *via* sensation to an image of movement, that the touch-organ has a large enough sensitive surface and is so mobile that one part can touch another so that we feel each one of these mutually touching points by way of the other. [89] If we move the finger-tip of one hand against the volar surface of the other hand, the finger-tip in its sensitive areas receives a large number of successive impacts from the uneven surface of the other hand, while at the same time more and more neighbouring regions of the other hand receive successive impressions from the finger-tip. We can describe a path with one finger-tip across the sensitive surface of the other hand and thus learn to discriminate and perceive sensations in one hand by way of the other hand; and, conversely, we also learn what kind of efforts must be made in order to move the finger-tip through predetermined directions and angles.

PERCEPTION OF SHAPE AND DISTANCE OF OBJECTS BY MEANS OF INTENTIONAL LIMB-MOVEMENTS

If we have learnt how to move our limbs intentionally in determined directions and angles, we have gained a new medium for obtaining information about the shape and distance of objects. By sensing the effort made by those muscles necessary to bring a limb to a given position or to hold it there, we gradually come to be aware continuously of the position of our limbs. If, for instance, the hands be held behind the back, so that they cannot be seen, and we allow somebody else to place them in positions where they are not touching each other, we still know at every moment just where they are. But when we are capable of moving our limbs intentionally and consciously, we can, by the movements we must make with our hands about the resistant object, build up a picture of the shape and size of the object. If we feel the upper surface of a table top with one finger of one hand, while the same finger of one hand, while the same finger of the other hand feels the bottom surface, we are capable of asserting, with eyes closed, how thick the table top is. It is difficult to conceive just to what extent our perceptions of the form of objects, and of the detailed form of their surfaces (coarseness or smoothness), depend on intentional movements of our limbs; our perception of the hardness, softness, and distances between objects, also depend on those. [90] Let one hand rest well supported, and close the eyes. If another person now brings glass, metal, paper, leather and other objects into contact with the finger-tips and rubs them against the fingers the objects can be confused, although one would have been able to distinguish them by moving the hand. A smooth glass plate pressed at first weakly, then strongly, then weakly against the finger-tip seems to have a convex surface: but if it is pressed at first strongly, then weakly, then strongly against the finger-tip, it seems to have a concave surface. It is interesting that if somebody pulls lightly at a lock of our hair, we can very accurately give the direction in which it was pulled: but with eyes closed, we cannot tell in what direction a knitting needle lies when it is pressed against a stationary part of our body: we cannot say whether it forms a right angle or other angle against the surface of the body. I have undertaken a more detailed examination of this problem, and have found that we do not feel the direction in which our hair is pulled directly, but that we resist with our muscles the displacement of the head and skin of the head initiated by a pull on the hair: and that we know by experience the direction to move the head or the skin thereof in order to resist such a movement. The correctness of this explanation may be vouched for by the following experiment. The ability to determine the direction in which our hair is being pulled is severely limited if our head is at the same time held still by another person: and is completely absent when in addition displacement of the skin is prevented as the hair is being pulled. For example the fingers can press against the skin of the head in front of and behind the hairs being pulled: and under these circumstances no movement of the head or the skin is elicited by the pull on the hair, and we therefore have no occasion to offer resistance to such elicited movement. [91] If one crosses two of the four fingers of one hand, and then lightly touches an object whose edges simultaneously touch the fingers, and which normally would have touched opposite sides of their fingers, the object appears to feel double: we think we can feel two metal balls being rolled between the fingers, or feel two tips or our nose, or two table-edges. In explaining this remarkable phenomenon the following must be taken into account. With two parts of an organ of touch we always have a double sensation: the two sensations we receive never fuse to one. But we are also led to the conclusion that the two sensations we receive are caused by two surfaces of one and the same object, as when for example, we hold a die between two fingers. If one of the surfaces is smooth and the other rough, or one warm and the other cold, then these sensations do not fuse nor do we ever imagine we are touching one and the same surface of the die. We assume that the four sensations we have when we touch the edge of a table with four fingers arise from the same surface angle, for we know from the position of our finger-tips that the sites of the four contacts lie along the same line. We moreover assume that we hold one and the same pencil between our fingers when we know the pencil entirely fills the space, and that we can press the pencil with one finger against the other finger and therefore assume the cause of the two sensations to lie in one and the same area of space. This is impossible to do if the two edges of two fingers touch the same object, the fingers themselves only being brought into such a position with some difficulty: normally they lie apart from each other and therefore cannot touch the same object. The correctness of the above account may be seen from the fact that, with the thumb and little finger, touched objects do not appear to be double. The thumb and the little finger, because they are highly mobile, can in fact easily and effortlessly be moved *via* their muscles to a position in which the otherwise diverted surfaces of the two fingers can touch the same object, and because, therefore, the movement of these two fingers is taken into account when forming our judgement, we perceive that it is one and the same object with which they are both in contact. [92]

II. The Sense of Pressure

By way of the localization-sense in the skin, supported by our own intentional movements, we can perceive *spatial relationships* in objects and detect their shape, size, distances from each other, roughness, smoothness, hardness and softness. Similarly, by way of the pressure sense in the skin, particularly when supported by our own intentional movement, we can recognise our own energy of movement and the forces which objects exert in resistance thereto. It is a fact that we prefer to use our eyes to recognise spatial relationships and thereby do so more comprehensively: but the sensations of the effects of our own forces and the forces of other objects derive entirely from the sense of touch, which might therefore be considered to be essentially a *sense of force*. Our concepts of *force* would be very much less well developed were we unable to feel pressure, or to sense competing forces in which an equilibrium is established so that no movements are produced, yet in which the forces can still be felt.

Forces actually exert their effects in two ways, by producing movements or by suspending movement, the latter giving rise to pressure or tension. Movements which do result may also be perceived visually, and this will also lead to our conceptions of force; but we do not feel movement directly but only imagine it for ourselves. On the other hand we do have sensations of pressure and its various degrees directly. But the development of our concept of force is best brought about by the fact that we ourselves can produce, by a greater or lesser exertion of will, pressure of a given degree, and are able to press one part of a touch-organ against another. While it is true that we are aware of our voluntary activity and the degree of effort it requires, we also feel, in addition, the effects of our voluntary activity, namely the pressure and the counter-pressure resulting from those organs of the touch-sense which are being pressed together. [93] In which other of our sense-organs do we possess a similar ability, that is, to produce an impression with the self-same sense-organ as is used for its perception? Where better than here do we have such an opportunity to be as aware of causative connections as when we become conscious of the execution of will, so that when we press one of our hands against the other, we feel both the pressure in the pressed hand, and the counter-pressure in the hand actually doing the pressing? In this respect the ability of the touch-sense, by way of which we feel pressure, is markedly different from the ability of the same sense by way of which we perceive warmth and cold: the concept of force as yielded by the sensations of warmth and cold is far inferior. If we could voluntarily develop heat and cold in a section of the skin and communicate it to another section of the skin, then indeed sensations from the touch-organ would give us a pronounced image of force. *Force* is the unrecognised cause of reciprocal actions between bodies, expressed in movements of pressure, but it is by no means a phenomenon to us and therefore we cannot tell if it is changeable or not. The only case in which we do know something more about this unknown source is precisely that case in which our will is the source or part of the source of a perceived pressure. For even if this pressure was partly created by a mechanism in our body, the will itself must provide the impulse for this mechanism, releasing it, as it were.

Physicists and chemists test the instruments they use in order to determine how reliable they are: for instance they test the scales with which they weigh and physiologists and anatomists test their microscopes in order to know their magnification. It is just as important for human beings to test their inborn media* of

* German, *Instrumente*.

sensation. I have recently undertaken such an examination for the touch-sense. In order to establish the precision with which we are capable of distinguishing various degrees of pressure, we can repeatedly place two differing weights of equal shape and surface-area on to the same part of the hand of an observer, the latter having his eyes diverted at the time. [94] We first place one weight on his hand, and then after removing it, quickly place the other thereon, then the first, and so on until the observer has come to an estimate as to which of the two weights is the heavier. My experiments have shown that this method had advantages over a method whereby two differing weights are put simultaneously one into each hand: for two simultaneous touch sensations cannot be so well compared as two successive. A series of experiments has shown that two weights can be most accurately compared if they are placed successively onto the same region of the same hand. It is slightly less advantageous to put one weight first on one hand, then remove it and place the comparison-weight onto the other hand. For the one sensation disturbs the other, for the two sensations mingle in a manner similar to that of two simultaneous tones, whose distance apart on the scale is less easily judged than if the two tones follow each other in rapid succession. This mingling of two simultaneous sensations is even more marked for smell than it is for touch or audition: it is extraordinarily difficult to compare two smells presented simultaneously, one to each nostril, from separate smelling-bottles.

This mingling of simultaneous sensations is an interesting phenomenon, but it is an even more interesting fact that a sensation presented earlier and therefore only present in the form of a remembered image, can be so exactly compared with a sensation actually being presented at the time. This is the case in the method of comparison of two weights described above, which has shown itself to be the most advantageous of all the methods. We would expect that the sensation presently available, the pressure we now feel, would always be so much stronger than the image formed of the pressure felt earlier, that the two impressions could not be comparable: but, as mentioned, this is absolutely and utterly not the case. Using various subjects, I have carried out series of experiments on the degree of incompleteness of the comparisons between two sensations presented with 2, 5, 10, 15, 20, 25, 30, 35, 40, and more seconds between the second sensation and the first, with which the second is to be compared. [95] In many persons the comparison was very poor after only 10 seconds. With large weight-differences even longer times than is the case for small weight-differences may elapse before the subject fails to distinguish the heavier from the lighter weight. I succeeded in distinguishing a weight of 14 oz., or sometimes even 14½ oz., from a 15 oz. weight when 15 or 30 seconds elapsed between the first and the second sensations. Even after 35 seconds I was able at times to distinguish the heavier from the lighter weight, but never after 40 seconds. If the difference between the two weights was greater, 60 to 90 seconds could elapse between the first and second sensation and I could still distinguish the heavier from the lighter weight, e.g. if the weights were in a ratio 4:5. With even greater weight differences even more than 100 seconds could elapse. I have made similar observations on visual sensations. I held pieces of paper marked with a black line in ink before the subject's eye and, after removing them, allowed 30 or 70 seconds to pass before I presented a second line which was similar to the first apart from being 1/11 longer: the longer line could still be distinguished from the shorter line. Discrimination decreased so sharply after a 70 second interval, however, that it was clear that the discrimination was impossible after 80 seconds. When the length of the lines were in the ratio 20:21 and therefore differed by 1/21, the longer line could be distinguished from the shorter even after 30 seconds, but not after 40 seconds. If the lengths of the lines were in the ratio 50:51¼ the longer line was distinguished from the shorter after 3 seconds, but not after 5 or 10 seconds. In this way one can measure and quantitatively express the clarity of the memory for sensations as it decreases from second to second. As we rarely have the opportunity of measuring such mental processes, I commend these experiments to the attention of psychologists.

It is of great interest to students of the touch-sense and of common sensibility that *apart from the touch organs we possess a second class of organs which provides us with a concept of the extent of the pressure or tension exerted upon our body: the voluntary muscles, along with the nerves and their central organs.* [96] We can judge the extent of a weight, or the extent of the resistance to be overcome by limb-movement, from the feelings of exertion of the muscles as we lift a weight, and also by the feelings of tension in the muscles if a weight pulls on a relaxed limb, thereby stretching them. It is important to investigate what the touch-sense alone, unsupported by muscles, can tell us in this respect and also what the muscles alone can tell us when they are unsupported by the touch-sense in the skin. Moreover we need to establish how this perception of weight is elaborated when both types of apparatus are used together during observations.

In my earlier investigation of the sense of touch I was not able to find a method for determining the contributions of the muscles alone to such judgements, but I have now found a very simple means for examining the contributions of either skin or muscle alone.

Let the observer rest his hand on a table so that it

is completely supported: then, while he turns his eyes away, place two different weights alternately on the two final sections of two or three selected fingers either on their dorsal or volar surfaces. If we alternately add and remove each of the two weights until the observer has arrived at a judgement of which is the heavier weight, his judgement will be based on sensations provided by the touch-sense of the skin. For there are no muscles in the two final sections of the fingers, and as those muscles which affect these regions by stretching from a distance are quiescent, muscles do not contribute to the sensations which arise, nor do the weights exert pressure on them.

But now let the observer grasp with his hand the joined corners of a folded cloth in which a weight is hanging, holding it either with arm bent or outstretched: then, taking the cloth from his hand, give him a second cloth enclosing a different weight. [97] Repeat this operation, without his being able to see the cloths, as often as is needed for him to arrive at a judgement as to which weight is the heavier. This judgement is based on the common sensibility in the muscles and not on the touch-sense in the skin. The cloth held in the hand does indeed rub against the skin, but exerts no pressure thereon. If the observer grips the cloth rather more firmly than necessary, so that it does not slip from his hand, he cannot arrive at any conclusion concerning the degree of weight merely from the pressure which the hand must exert to hold the cloth so that it does not slip. Ten persons, half of whom were male, compared 78 and 80 ounces by lifting the weights in cloths as described: only two failed to distinguish the heavier from the lighter weight, while seven persons correctly determined, each time in a series of three experiments, which was heavier. In some instances, four to seven experiments were undertaken, and in every case the weight was judged correctly. One of the 10 observers judged correctly seven times out of eight, and once wrongly. *It must therefore be assumed that the majority of persons, without previous lengthy practice, can distinguish between two weights, which bear the ratio 39:40, using the common sensibility in the muscles.*

Accuracy in weight-discrimination, based entirely on the touch-sense of the skin, is not quite as well developed. According to my earlier investigations, which I have confirmed in more recent experiments, *weights bearing a ratio of 14½:15, i.e. 29:30, can be distinguished only with the greatest difficulty* if the weights are placed successively on the same sections of the fingers resting on a table.

If the two methods are combined, discrimination is, to say the least, no better than the method of lifting fairly large weights successively: we therefore cannot assert that the two methods combined would give much better results than would be achieved by lifting only fairly large weights. [98] But for smaller weights a combination of both methods of investigation seems on the whole to be useful.

If weights in the ratio 29:30 placed successively on the fingers are to be distinguished accurately, several precautions should be followed. They must be equally warm, or surrounded by a poor heat-conductor, since cold metals seem to be heavier than warm ones.[41] They must always be placed on the same part of the skin; they must have an equal base-area in contact with the skin; and, when one weight is removed, the other must be quickly placed in its stead, but not dropped there suddenly.

We saw earlier that the localization-sense in man is more than 50 times more acute in the tip of the tongue than it is in the middle of the back, it is 7 to 8 times more acute on the palm side of the finger-tip than on the back of the hand, 10 times more acute than on the forehead, 18 times more acute than on the middle of the lower arm, 20 to 36 times more acute than on the middle of the upper arm if the latter be touched transversely, and about 50 times more acute than on the middle of the upper arm or thigh if the latter be touched longitudinally. We may therefore ask whether the acuity of the touch-sense insofar as it enables us to perceive small differences in pressure exerted on our skin and to sense small differences in weight, is the same for other parts of the skin as well. The answer is definitely no: and, as we shall see later, the same is also true of the ability to distinguish between small differences in temperature.

Lesser degrees of pressure can indeed be better distinguished with the fingers than with the skin of the lower arm: and this ability so even less well developed on the human back or abdomen than it is on the lower arm. But these differences are much less noticeable than is the case for analogous differences in terms of the acuity of the localization-sense, for the total difference is such that the acuity of the touch-sense in the middle of the lower arm is only 6, whereas on the fingers it is 7. [99] This comparison suggests that we can investigate the acuity of the touch-sense in two ways. Firstly, weights can be placed simultaneously on the fingers and on the lower arm: the weight lying on the lower arm seems lighter than that on the fingers. In my work on the touch-sense I have reported numerous experiments and found, as previously mentioned, that the ability to sense the pressure of a weight and to accurately perceive differences between weights was only slightly different in the two last sections of the fingers from that on the skin in the middle of the lower arm. For example, if 5 ounces were placed on the volar side of the three middle fingers, and 4 ounces on the middle of the lower arm, these body parts resting on a table, a stronger pressure was felt on the fingers. But if 4 ounces were placed on the fingers and 5 ounces on the forearm, the pressure seemed about equal, and remained so until the weights on the lower arm had been increased to 7 ounces; that on the fingers staying at 4 ounces. Only then was the sensation of pressure decidedly greater on the lower arm. The strength of feeling in the fingers as compared with the middle of the lower arm therefore was in a ratio of about 7:6, or more accurately, 1.183:1. On the other hand, the developed localization-sense in these regions was in a ratio of about 9:1, that is, it was 9 times greater in the middle of the finger than on the middle of the lower arm. I have described 15 experiments carried out on various body regions all of which have given similar results.

But there is a second method of investigating the accuracy of the ability to sense pressure, namely by successively placing two weights, differing only slightly, upon the same fingers, and observing how small a weight-difference can be discriminated: and then carrying out the same experiment on the middle of the lower arm in order to find out whether the weight-difference must be greater here than on the fingers if the difference is to be perceptible. [100] The results of these experiments, which are more accurate and reliable than those just described, showed that the same observer could distinguish the heavier from the lighter weight, on the fingers, when the weights stood in a ratio of 20:19.2; whereas on the middle of the lower arm such a small difference was not recognized: the weights had to stand in a ratio of at least 20:18.2 in order for the heavier to be distinguished fromthe lighter.

In the same way the fingers and the forehead were compared in this respect: it was found that weights laid on the forehead had to stand in a ratio of at least 20:18.7 in order for the heavier to be discriminated from the lighter, while on the stationary fingers weights could be distinguished if they stood in a 20:19, or even 30:29 ratio.

It has already been shown that there is no contradiction in the fact that regions in which the localization-sense is highly variable possess only slight variability in the ability to discriminate weight-differences. This is because the development of the localization-sense seems to be based on particular mechanisms which are not involved in the development of the ability to discriminate weights. In order for the localization-sense to be highly acute it is requisite that on a region of skin of a given area very many elementary nerve-fibres must end in a certain order: but if the ability to perceive differences in weight or temperature is to be appropriately developed, it is only necessary that many sensitive points lie within an equally large surface area of the skin. Whether these points owe their sensitivity to one and the same elementary nerve-fibre, or to several separate nerve-fibres seems to be of no, or at least very little, relevance; for if impressions are made on the same nerve-fibre at several points, it then seems that the overall impression upon the whole fibre is thereby strengthened.

III. The Temperature Sense

The sensations of warmth and cold are not like the sensations of brightness and darkness, for the former are positive and negative quantities between which lies a null point determined by the source of heat within us. [101] If objects surrounding and touching our skin have a temperature such that the temperature of our skin, quite apart from the fact that we have a source of heat within us, neither rises nor falls, then these objects themselves appear to be neither warm nor cold. If their temperature raises that of the skin, they appear to us warm: but if they cause the skin temperature to fall, we say that they feel cold. But absolute darkness is the null point of brightness, and the various degrees of illumination, from darkness through to the most brilliant brightness, are therefore all positive quantities.

A thermometer shows the temperature of mercury at all times, whether it be either rising or falling, or staying steady. It is a different matter where the touch-sense is concerned. It appears here that we can perceive the process of rising or falling of the skin-temperature rather than the degree to which it actually has risen or fallen. We do not feel, for example, whether our forehead is warmer than our hand until we actually touch our forehead with our hand, at which point we often perceive a large difference: sometimes our hand is warmer, sometimes our forehead. If we lay our hand on our forehead, the colder of these parts lowers the temperature of the warmer, and *vice versa*, and we feel this raising and lowering of temperature in both parts. Without this reciprocal effect we would not be able to compare temperatures in the various parts of the skin. Therefore we confuse a rapid lowering of skin temperature with a large drop in the temperature of the skin. If we immerse one hand in moderately cold water, while repeatedly but only momentarily dipping the other hand into the same, we imagine that the latter gives rise to a more pronounced degree of coldness than does the former; yet the temperature in the skin of the former falls to a lower level than that of the latter. When the latter is not being immersed there is no warmth being lost, in fact, part of the lost heat is replaced by its own inner warmth.

At first sight the following experiment may appear to contradict the above viewpoint. If part of the skin of the face, e.g. the forehead, is placed in contact with a metal rod at $+2^{o}$R for a short time, e.g. 30 seconds, and the rod is then removed, the cold can be felt to persist in that region of the skin for a further 21 seconds or so. [102] In view of what has just been said, it would have been expected that we would have had a sensation of warmth, as the cooled part of the skin became warm again. I suggest therefore, that in this latter case, the feeling of cold does not arise because the nerves of the cooled skin region produce the sensation of cold, but because the nerves of the neighbouring regions of skin to which cold is now transmitted from the cooled region, produce the sensation.

We are as little acquainted with the apparatus at the ends of the touch-nerves which transmits the sensations of warmth and cold to these nerves as we are with that of any other sense-organ. But we shall, we hope, come to understand it in the future after further microscopic study. *It is therefore still uncertain, whether those structures responsible for sensations of pressure are the same as the ones that transmit sensations of warmth and cold, or whether there is a structure which is specialized to transmit these sensations.* Unquestionably the latter's functioning would be based on the fact that the volume of a body increases if its temperature rises, and decreases if its temperature falls. This kind of change, following known physical laws, will affect fluids to a much greater extent than solid substances. The soft, fluid-filled extensible cells of the tissue which forms the sensory papillae must on account of the large amount of liquid which they contain expand far more under heat, and contract far more under cold, than does the dry skin which surrounds the skin papillae like a sheath. It can be conceived therefore that temperature changes will give rise to pressures and tensions between the skin papillae and their sheaths. Many hypotheses as to how pressure or tension could affect nerves could still be based on those characteristics of fluid bodies which enable them to be more greatly increased in volume by temperature changes than are solid bodies.

As a starting point for a further examination of such structures, it must be noted, as was shown experimentally above, that cold bodies seem to be heavier than warm ones of equal weight, and that therefore cold appears to act like pressure, and the two types of sensation, if evoked simultaneously, are confused. [103] This experiment thus permits the view that even sensations of warmth and cold rest upon pressures and tensions being exerted on the nerves.

It will be seen from the above that we do not always perceive the temperature of bodies correctly by means of the touch-sense. Cold bodies which are also good conductors of heat, seem to be much colder than others of the same temperature but poorer conductors. A cold piece of wood seems to be much less cold than an equally cold metal rod: water seems colder than oil, when both are the same temperature: this is true of all good and poor conductors, if they are warmer than our blood. A good conductor withdraws heat more quickly from our skin in the first instance, and transfers heat to it more quickly in the last instance. Another example is that warm fluids seem warmer, and cold fluids colder, if an immersed hand is moved about in it. If we keep our hand still, the liquid in contact with it takes on a different temperature, closer to that of the hand; if we move the hand this fluid is driven away from the hand and new liquid, whose temperature has not yet changed, continually come into contact with it. The effects of draughts, which often produce unpleasant colds, are based on this.

A second reason for our not always correctly perceiving the temperature of objects is that the skin itself does not always preserve the same temperature, as e.g. when there is less blood than usual flowing to a certain region of the skin, or when the sustained application of moderate cold makes the skin itself colder. In these cases a new state of equilibrium is gradually attained in which the cooled part of the skin releases just as much heat as is being conducted to it from inside. Objects which are warmer than the skin, and thereby transfer heat to it, now seem warm even if they have a lower temperature than that usually possessed by the skin, and would therefore normally feel cold. [104] A doctor wishing to test a patient's temperature accurately must therefore take pains to keep his hands at a constant temperature.

If I immerse my hand for one minute in water of a temperature of +12.5°C, and then put it into water of 18°C, I experience a feeling of warmth in the latter for a few seconds; but gradually a feeling of cold sets in, this persisting for as long as the hand remains immersed. The rise in the temperature of the cooled skin therefore gives a feeling of warmth even when the temperature thus arising is such that it should still be felt as cold. But this feeling of warmth only persists for as long as the temperature takes a rise: afterwards, cold is felt, for more warmth is taken from the skin by the water than is conducted to it from the inside.

Water at a temperature of 35°C (28°R), and therefore 2.5°C (2°R) cooler than the blood (whose temperature I shall here takes as +37.5°C or +30°R) gives rise to a feeling of warmth providing the hand is less warm, as is usual at room-temperature, and is not immersed for very long. If, however, the hand is immersed for a long time, a slightly cold sensation results. From the outset, warmth from the water is transmitted to the hand, but once the temperature of the skin on the hand has risen due to warmth from both the interior of the hand and the exterior, a movement of the hand in the water cools the skin down, evoking a slight sensation of cold.

Water at a temperature of +36.2°C (29°R) always gives rise to a feeling of warmth in the immersed hand. Admittedly, this temperature is somewhat lower than that of the blood; nevertheless, it appears to cause an increase in the skin-temperature, possibly because the skin, when brought into contact with fluid of this temperature, does not lose as much heat outwardly as is produced by its own internal sources of heat.

The ability to feel cold and warmth is not equally developed over the whole surface of the skin; variations in this capacity are however, as noted earlier, much less marked than those involving the localization-sense. [105] That the temperatures of objects touching us are more quickly perceived on some parts of the body than on others should not be confused with the fact that our accuracy in perceiving very slight differences in temperature is better developed in some areas than others. The first of these phenomena is due to the fact that the epidermis is thinner, because the epidermis is itself insensitive and a poor conductor of heat. So the thicker it is, the longer it will take before the heat or cold penetrates to the sensitive parts of the touch-organ, and the more the latter is capable of being placed on contact with hot bodies without a burning sensation arising. A good opportunity for distinguishing between the variations in sensations of heat and cold arising from these two separate sources is provided by placing both hands in a deep vessel of cold or warm water in such a way that the palms face each other without touching. If the temperature of the water is e.g. $+1.2^{o}C$ ($+1^{o}R$) or $+2.5^{o}C$ ($+2^{o}R$), then the cold is first felt more strongly on the backs of the two two hands than on the palms because the epidermis on the back of the hand is much thinner. But after about 8 seconds the cold felt in the palms of the hands starts to dominate, the sensation increasing to a point at which it is soon quite beyond doubt that the same water has evoked a much stronger feeling of cold in the palms than it has on the backs of the hands. The same is experienced if warm water is used, the corresponding feelings being those of warmth. Perhaps my observation that warm and cold evoke rather stronger sensations in the left hand than in the right is to be explained by the fact that the epidermis in the left hand is rather thinner than in the right. It is therefore certain that, just as the localization-sense in the palm is more developed than that on the back of the hand, so the temperature-sense is also better developed on the palm: however it is equally certain that this last difference is minor as compared with the large difference in localization-sense between the palm and the back of the hand. The reason for this also seems to lie in the fact that an accurate temperature-sense demands a large number of points sensitive to temperature, but that it is not necessary that a specific elementary nerve-fibre actually end in each such sensitive point: it suffices if the same elementary nerve-fibres, by virtue of branching, or twisting a great number of times, sensitize a large number of points in the skin. [106] By contrast, the acuity of the localization-sense is due to many elementary nerve-fibres ending in a definite order at adjacent points in the skin.[42]

It is of the greatest interest that the size of the portion of skin simultaneously stimulated by a warm or cold object also influences the experienced sensation of heat. If into the same hot or cold fluid the index-finger of one hand be immersed at the same time as the whole of the other hand, the feeling in the two extremities is not the same: that of the whole hand is stronger. This increase in the strength of sensation, based on the fact that the same influence is exerted simultaneously on many more sensitive points, is easily confused with the strength of the impression evoked under other circumstances by the degree of the temperature itself. So cold water is perceived as colder, and warm water, warmer, by the whole hand than by a single finger, quite apart from the fact that one knows the two extremities to be immersed in the same water. If one does not know this, and dips both extremities into containers filler with water of an unknown temperature, then one is led to believe that, if a whole hand is placed in water of +29.5°R, and a single finger in water of +32°R, the former is warmer. A similar illusion is experienced if water of +17° and +19°R is used, the water at +19°R, experienced by the whole hand, feeling colder than the other despite its actually being 2°R warmer. *The sensations of heat transmitted to the brain from many sensitive points would therefore appear to summate to produce a single total impression.* [107] As we will show later, this is also the case when hot and cold impressions produce pain, and this is not simply because the hot or cold reaches a particularly high level, but because the impressions of heat and cold affect a large portion of the skin. This summation of impressions, whereby a greater and even painful total impression might be produced, is one of the phenomena which make it likely that the brain is the place in which the movements excited in the nerves emerge into consciousness. *The closer together the regions of skin simultaneously excited, and probably, the closer together the regions of the brain to which the impressions are conducted, the more easily the sensations fuse into one; the further apart these regions, the less easily do they fuse.*

If we simultaneously place two fingers of the same hand, e.g. the thumb and index-finger, one into each of two adjacent containers filled with water of differing temperatures, the two impressions do not fuse into one, however our ability to compare the two temperatures is severely impaired because of their close proximity. We are much less confused if we simultaneously dip the two thumbs of the two hands into the two containers. But even here there is some difficulty in making the judgement, and we are therefore most adept at comparing two temperatures if we alternately dip the two thumbs into the two containers, and the comparison is made most easily of all if we dip the same finger or the same hand first into one container, and then into the other. *Under these conditions it is possible, with a great deal of concentration, to detect with the whole hand temperature-differences as small as 1/5 or 1/6 of a degree on the Réaumur scale.*

Most persons can accurately detect a difference of 2/5 of a degree. We might suspect that the perception of such a small temperature difference would only be possible for temperatures close to blood temperature. [108] I must note however that greater differences were not needed to distinguish between two temperatures near 14°R than were necessary to distinguish between two temperatures near that of the blood.

As noted earlier, it is very difficult to determine whether a temperature-difference between two objects on one area of skin can be more clearly detected than on another because the skin is more sensitive or because the epidermis is thinner. Accordingly, it is very important for physiologists to recognize that *sensitivity to temperature changes related to these two factors is itself very variable even in closely adjacent regions of skin; and that therefore regions of skin which are more sensitive to temperature are interspersed with regions which are less sensitive in this regard.* If objects of a fairly constant temperature are brought alternately into contact with various parts of the skin, it may be observed that the sensation thereby aroused is very much more marked for some areas than for others. In order to keep objects at the same even temperature for touching the skin in such experiments, I filled two very elongated glass phials with oil and introduced thermometers into the phials through holes bored in the stoppers. If I now heated or cooled the phials in water, and then wiped them dry, the thermometers then showed their temperatures. It might even have been better to use mercury instead of oil. The skin of the face seems to be more sensitive to temperature than any other part of the skin. The eye-lids and cheeks are particularly sensitive to warmth and cold. The lips, which have a much finer localization-sense than these areas, are less well developed in their sensitivity to heat and cold. The localization-sense, as mentioned earlier, is most acute in the middle region of the upper lip, decreasing steadily outwards, especially toward the cheek; sensitivity to heat and cold on the other hand, is more acute at the sides of the upper lip, most acute on the cheeks, but less so in the centre of the upper lip. In order to repeat these experiments with an object of small surface area, I took a large heavy door key with a solid shank and a large heavy bit, and first dipped it into a large amount of mercury of a given temperature, then laid it for some time on a very cold stone slab by the window, the temperature there being shown on a thermometer. [109] After the key had attained the temperature of the slab, I repeatedly and alternately placed the rounded end of the key against two regions of the skin, the temperature sensitivities of which I wished to compare. Using this method, I not only obtained the same results as I had with the phials, but was able to compare even smaller regions of the skin with one another. These experiments showed that the eye-lids were more sensitive at their inner and outer corners than in the middle, and that the tip of the tongue is one of the most sensitive areas. Sensitivity to temperature-changes, according to tests I carried out on myself, is much greater on the face than on the neck. The skin areas separating the two sides of the face, chest, abdomen and back, were less sensitive to temperature variations than were the areas to the sides. For example, sensitivity to temperature-change is much coarser at the tip of the nose than at the side of the tip of the nose; it increased markedly at the sides of the nose, being greatest at the lower edge of the outermost parts of the sides of the nose. It is greater on the cheeks and near the tragus of the ear than on the lips: on the lower edge of the jawbone than on the chin; on the temporal region just above the zygomatic arch than on the middle of the forehead, above the glabella. The insensitivity to heat and cold of the inner skin of the nose was most surprising as compared with the higher excitability to such impressions shown by the skin of the auditory canal.

The trunk and the extremities are also sensitive to heat and cold to different degrees, which may depend partly on the greater thinness of the epidermis, partly on the arrangement of the sensitive regions of the skin. For example, the first joint of my index-finger in the palm, the part below the gap between the fingers, is more sensitive to heat and cold than the same regions in the third, fourth, and fifth fingers. [110] The ball of the thumb shows greater sensitivity than does the ball of the little finger, the region of the elbow near the olecranon is more sensitive than the skin on the middle of the biceps or the triceps, and the region of the trochanter major is more sensitive than is the middle of the region of the crista ilei. Even if these observations are of little practical application, at the moment, they might be useful in times to come in helping us locate the microscopic organs sensitive to temperature, and to distinguish them from the organs of localization-sense and perhaps of pressure-sense, as well. On those regions of the skin which are highly sensitive to warmth and cold, the contact of very hot or cold objects causes pain more quickly than is the case in less sensitive areas, provided the thickness of the epidermis does not prevent it. We shall say more on this when we come to discuss common sensibility.

DO TWO SENSATIONS ARISE WHEN TWO TOUCH ORGANS COME IN CONTACT?

We can bring two regions of skin into mutual contact and thereby cause one of the regions to cause an impression of pressure, warmth, or cold in the other. In other sense-organs this is not possible: we cannot, for instance, look with one eye into the other. The question is therefore raised whether the two impressions arising simultaneously when the two touch-organs are brought into contact blend into a single sensation, or whether they remain separate: and whether, in the latter case, we can ourselves determine, by deliberate control or concentration of attention, which of the two impressions should enter our consciousness, or what the circumstances are which determine which of the two sensations will enter our consciousness.

Experiments I have carried out prove that the impressions do not fuse to a single sensation. If for instance we bring a cold limb into contact with a warmer one, we do not feel an intermediate temperature, but, under some circumstances, cold, under others, warmth, and occasionally, cold and warmth alternately. [111] When the sensations of cold and warmth alternate rapidly, we think we perceive warmth or cold simultaneously, or one after the other; but we are incapable of conceiving sensations of cold and warmth as fused into one, as we are able to imagine a high and a low musical note as one if we think of them in terms of an interval.

But which of the two impressions reaches consciousness depends only very slightly upon the degree to which we concentrate; usually this is determined by other circumstances. If we close our hand for some time, its temperature thereby rising nearer to that of the blood, and then open the hand to touch the forehead lightly, then we feel the warmth of the hand on the forehead, but we do not feel the cold of the forehead on the hand. But if we pay special attention to the question of which of the two is felt to be the object, we find that we perceive the forehead as the object touched with the hand, and never the hand as the object touched by the forehead. This unexpected result, which appears to be something of a conundrum, can be explained in the following way. The forehead has a thinner epidermis than the palm, and the warmth of the palm therefore penetrates more rapidly to that region of the skin in the forehead supplied with touch sensitivity than is the case *vice versa*. The attention is thereby focused on the more rapid and intense sensations of temperature arising from the former. On the other hand, the palm is endowed with a more developed localization-sense than is the forehead, so that our attention is focused on the hand, in which the sensations of pressure are stronger and more definite. Moreover, all things being equal, we direct our attention to that part of the body which we move, and, we therefore perceive the immobile region to be the object touched by the region voluntarily moved. These two phenomena come into play on the forehead and cause us to feel the forehead as the object touched by the hand. Indeed, by concentrating, we can gradually feel the outstretched fingers as being on the forehead, but only to about the same degree as if we were to lay the outstretched hand on a cold table and sense the shape and position of the individual fingers on the table more clearly than if they were merely in contact with the air. [112] This greater sensitivity in the fingers exists because we feel that the parts of the table in contact with the hand must themselves be sensitive.

In the case of small temperature differences, of the kind we have thus far been discussing, we find that they are no better perceived with warmer regions than with colder, or *vice versa*. If one hand is a little warmer than the forehead, and the other a little colder, then, if we lay the warmer hand on the forehead, we feel warmth, and if we lay the colder hand on the forehead, we feel cold: that is, we feel the temperature of the hand on the forehead in both cases.

But it is otherwise if we artificially expose the forehead to a markedly warmer or colder temperature, e.g. by applying a towel or other objects of room-temperature (say 18°C) to the forehead, cooling it by applying the objects or cool parts of them. If we now open our hand, hitherto closed tight, and lay it on the forehead, we first feel cold, then warmth: and eventually, some parts of both hand and forehead feel warm, while others feel cold. Under these conditions we therefore first feel the artificially induced cold temperature of the forehead with the warm hand. If the same experiment is carried out on parts of the body which are essentially the same, e.g. the two hands with their volar sides laid together, the one however having previously been cooled by laying it on a table at 18°C enclosing it in a cloth at that temperature, the other having been warmed by being tightly closed; then we first feel the unusual temperature, namely the cold, after that the warmth is felt, and eventually we feel various parts of the two hands, some of which appear to be warm and some to be cold.

If we dip one hand for some time into cold water at 17°C, dry it, and then grip the back of the other hand (kept at its normal temperature) the warmer hand is felt as an object, but it seems cold. The point is that the palm has a finer localization-sense than the back of the hand. Therefore the palm feels the back of the warmer hand as an object. [113] But the back of the warmer hand has a much thinner epidermis than the palm, and the cold therefore penetrates more rapidly into the former than the warmth does into the latter. We therefore feel the resulting temperature-change with the back of the hand, but imagine we feel it with the palm.

If one hand is dipped in warm water, and the two hands then laid together, the warmth of the formerly immersed hand is felt with the other hand. If one hand at its normal temperature is laid on the cooled forehead, the cold of the forehead is perceived. One therefore always feels the induced temperature of the skin in a part of the body at its normal temperature. It will be clear that very intense warming or cooling which, impairs or, as we have previously seen, cuts off skin sensitivity should be avoided; for in these cases, the warmed or cooled region evokes sensation only in the area touching it, and is itself incapable of receiving temperature stimuli. Despite the fact that in the above-mentioned experiments a feeling of warmth does not fuse with a feeling of cold to produce an intermediate temperature, the simultaneous presence of two opposing sensations does result in a noticeable disturbance; the feeling of cold or warm is therefore much clearer and more definite if somebody else puts his hand on our forehead than if we touch our forehead ourselves. If another person lays his hand on our forehead, we not only feel the temperature of the hand clearly, but also perceive it as an object. These experiments re-emphasize, moreover, that our ability to control our perceptual faculties and to focus them on sensations we wish to observe is more limited than is usually believed.

Volkmann has previously made the interesting observation, with respect to vision, that our ability to focus attention onto one visible object or another is considerably aided by physiological mechanisms and has even voiced doubts as to whether or not human volition would be able to control perception in the body without these mechanisms. It is indeed true that the mind is greatly facilitated, in indirectly controlling and directing perception while observing objects in the visual field, by the design of the eye. [114] Because only a very small part of the retina, about 1/3 or, at most ½ Paris *lines* in diameter, is organised in such a way that the objects forming images on it can be seen with sufficient clarity, and as it is only the centre of this area which provides the greatest visual acuity, we move our eyes in such a way that the object upon which we wish to focus has its image formed on this most sensitive region of the retina. If the objects simultaneously forming images on the retina were of about equal clarity, which would be the case if all parts of the retina were equally sensitive, it would unquestionably demand great mental effort to divert the attention from certain equally strong and clear sensations and direct it to others. It is much easier for us to move the eye or the head about and thereby cause only those objects we particularly wish to view to evoke intense and clear impressions on the eye. This impression becomes even stronger if we direct both eyes at the same object in such a way that the extended optical axes intersect at the object.

Another physiological mechanism, also demonstrated by Volkmann, lies in our ability to adjust the eye to varying distances, which enables us to see both close and distant objects clearly. We can even stare vaguely into space with the result that we do not clearly see objects near at hand. In this instance, our eyes are not focussed for the distance where visible objects can be found, but rather for one in which there is nothing to see. This is the case with persons we describe as being lost in thought or day-dreaming. Having said this, I would not wish to deny man's ability to focus his perceptual faculties directly as well. This ability is demonstrated in the experiment in which, keeping our eyes fixed in a forward direction, we can still focus the attention on an object to the side, such as a hand held out sideways level with the eye.

ON THE SMALLEST DIFFERENCES IN WEIGHT PERCEPTIBLE TO THE TOUCH SENSE, THE SMALLEST DIFFERENCES IN LENGTH OF LINES PERCEPTIBLE TO THE VISUAL SENSE, AND THE SMALLEST DIFFERENCES IN PITCH PERCEPTIBLE TO THE AUDITORY SENSE. [115]

It appears from my experiments that the smallest difference between two weights which we can distinguish by way of feeling changes in muscle-tension is that difference shown by two weights roughly bearing the relation of 39 to 40, i.e. when one is about 1/40 heavier than the other (see p. [97] above). If we go by the feeling of pressure exerted by the two weights on the skin, we can actually distinguish a weight-difference of 1/30, i.e. when the weights are in the relation 29 to 30.

If two lines are perceived one after the other, a person who is excellent at visual discrimination can, according to my experiments, detect a difference between two lines bearing a ratio of 50:51 or even 100:101 to each other. Persons with a lesser ability in this regard can distinguish lines differing in length by about 1/25. The smallest difference between the pitches of two tones almost in unison which can still be detected by an artist who hears them in sequence is, according to Delezenne,[43]* $\frac{1}{4}$ comma $(81/80)^{\frac{1}{4}}$.** A music-lover, according to this

---

* C. Delezenne (1776-1886), physicist who taught at Lille and did much research into the physics of music.

** A comma is a small musical interval, equal to about a quarter of a semitone. See note on p. 96.

author, can distinguish only ½ comma $(81/80)^{\frac{1}{2}}$. If the tones are heard simultaneously, such small differences in tone cannot be detected, according to Delezenne's experiments. ¼ comma represents a ratio of approximately 321:322; ½ comma represents approximately 160:161.

I have shown that results for weight-judgements are the same whether ounces or half-ounces be used: for the results do not depend on the number of grains making up the extra weight, but upon whether the extra weight is 1/30 or 1/50 of the weight to be compared with the second weight. It is the same with the comparisons of the lengths of two lines or the pitches of two tones. [116] It makes no difference whether lines of about 2 or 1 inches in length are compared, provided they are observed successively and not simultaneously and next to each other. Nonetheless, the difference in length of the two lines is twice as great for the 2 inch lines as it is for the 1 inch ones. Of course, if both lines are close together and parallel, then only the ends of the lines are inspected to see how much one line is longer than the other, with the determining factors being the degree to which one line exceeds the other, and the proximity of the lines to one another.

Even in comparing the pitch of two tones, it matters little whether the two tones are seven intervals apart, provided they do not lie at the extreme ends of the scale where exact discrimination of small differences in pitch becomes more difficult. Here again it is not the number of vibrations by which one tone exceeds the other which is important, but the relationship between the numbers of vibrations of the two tones to be compared. If the vibrations of the two tones were to be registered by the mind, we might imagine that it is only the number of vibrations by which one tone exceeds the other which is being recorded. If we first fixate one line visually and then another, and images of the two are formed successively on the most sensitive part of the retina, we should be inclined to assume that we are comparing the traces of the impression left by the first image with the impressions being made by the second image on the same part of the retina, and that we are noting how much the second image exceeded the first and *vice versa*. This is how we compare two rulers: we cover one with the other to see how much one of them overlaps the other. The fact that we do not use this otherwise so advantageous method implies that we cannot use it: and that therefore the preceding impression does not leave a trace in the retina or brain which can be compared with the succeeding impressions in the way suggested. That our minds operate in a different way when comparing the lengths of two lines can be seen from the fact that we are able to compare one line with another when the two lines are too long to form complete images simultaneously [117] on the most sensitive part of the retina. In these cases we have to move the eyes and thereby ensure that all the various parts of the same line form successive images on these regions of the retina. Under these conditions we therefore also take the movement of the eye into account and it is only by doing so that we form an idea of the length of the lines. If the impression of visual images we preserve in memory were actually traces left by sensory impressions in the brain, and whose spatial relationships corresponded to those of the sensory impressions, and thus were daguerreotypes of them, it would be very difficult to recall an object which was too large for a single image to be formed all at the same time on the sensitive parts of the retina. It does seem to me indeed that an object which can be comprehended at one grasp does impress itself better on our memory and imagination than does an object which can only be comprehended in a succession of glances, involving eye movements: however the former object can also be conceived by the imagination. But in this case it would appear that the conception of the whole object must needs be built up from the separate portions perceived all at once.

If two lines of 20 and 21 Paris *lines* are compared, then though the latter is about 1/20 longer, the absolute difference between the lengths is 1 Paris *line*: on the other hand if one compares two lengths of 1 Paris *line* and 1.05 Paris *lines*, the difference is still one of 1/20, but one line is only 1/20 Paris *line* longer than the other, and in this case the absolute difference is 20 times smaller than in the former. 1/20 Paris *line* however is a quantity bordering on the barely visible, like a fine pinhole. We are still capable of seeing a point whose diameter is 1/20 Paris *line*; and persons with very good visual discrimination are capable of distinguishing between the lengths of two lines, one of which is only 1/20 Paris *line* longer than the other. Two observers to whom I presented such lines were both capable of differentiating the longer from the shorter, indeed their visual discrimination even surpassed this. I myself distinguished two lines whose relative difference was 1/20, of which one was between 1/17 and 1/18 Paris *lines* longer than the other. The ability to grasp relationships between total quantities without measuring these quantities with a small unit, and without knowing the absolute difference between the two, is an exceptionally interesting psychological phenomenon. [118] In music we can perceive tonal relationships without knowing the number of vibrations involved: in architecture we can conceive of spatial relationships without having to measure them in inches: and in the same way we can perceive weight differences through sensory intensity or muscular exertion.

THE RELATIONSHIP OF THE TOUCH-SENSE TO OTHER SENSES

The touch-organs, like the visual organs, have a localization-sense, but to a far less developed degree: we therefore owe our accurate perception of spatial relationships to both senses.

The fact that rapidly successive impacts to the touch-organs give rise to a single fused sensation, but that the intervals between the impacts actually cause the sensation to change, suggest that we have here a transition between touch and hearing. We feel vibrations as a quivering, perceptible to the auditory organs as a tone, and this quivering can occur in the widest range of variations. This can be perceived well when skating: various changes in sensation are observed if we cross from very smooth to very rough ice, these being particulary noticed in stretches where the elevations and depressions in the surface lie at regular intervals, being large on one stretch and short on the other.

If the skin is covered with a very thin and moist epidermis and is at the same time sensitive, we have a *transition from the touch-sense to the senses of smell and taste*. This is the case on the conjuctiva of the eye, where vivid sensations are experienced if this tissue comes in contact with sulphuric acid or ammonia fumes. That part of the mucous membrane covering the floor of the nasal cavity, and the mucous membrane in parts of the throat, are also sensitive to fumes from ammonia or ether. If the ability to smell is eliminated for some time by filling the nose with water, the water then being removed (see above, pp. [28] and [52]) it is still possible for the floor of the nose, the palate, and the throat to sense ammonia, sulphuric acid and eau de Cologne fumes. [119] These are difficult to describe, but might be called 'stinging' in the case of ammonia. If ammonia fumes are breathed in orally from a wide opening in the mouth of a large bottle filled with corrosive ammonia, there is no feeling in the tongue, but there is a stinging sensation along a wide region of the throat. In smelling and tasting, sensations arising from the touch-organs intermingle with those arising from the organs of smell and taste.

COMMON SENSIBILITY
COENAESTHESIS*

The words 'common sensibility' are used by the majority of psychologists to designate our ability to perceive our own sensory states, e.g. pain; this condition is therefore distinguished from our ability to have a sensation derived from an object distinct from our own sensory state itself, e.g. the sensation of a colour or a tone. This ability has therefore never been taken to be a typical sense-modality. Rather, many have assumed that *all our sensory nerves*, under certain circumstances, can provide us with sensations of this kind, but that there are also sensory nerves which, because they are not associated with any particular kind of sensory organs, do not give us access to typical sensory perceptions, but rather only to sensations of 'common sensibility'. Some even believe that when we first use our senses, all impressions are sensed merely as variations in our own general sensory state, and that it is only gradually, by comparing and interpreting our sense-impressions, that we have come to learn to interpret certain sensations as objects.

Sensations of 'common sensibility' often occur with other sensations simultaneously, and then represent only various effects of one and the same impression, e.g. the nausea which can be evoked by a smell, or the pleasurable and unpleasurable aspects of sensations which are perceived directly and simultaneously along with the sensations themselves, and which do not arise as the result of a comparison between sensations. [120]

Thus, the name 'common sensibility' denotes an awareness of our own sensory state; it is provided by all parts of the body supplied with sensory nerves, except for the specific sensations also provided by many of these parts.[44]

This viewpoint has recently been subject to doubt, following experiments of Magendie described on p. [39]. These led him to conclude that physical damage to the nervous tissue of the eye, optic nerves, olfactory nerves, and auditory nerves of mammals caused no pain. Since it has been known for some time that physical damage to the cortical substance of the cerebrum and cerebellum or to their neighbouring layers of white substance, or even damage to the corpus callosum, cause no immediate signs of pain: and since the experiments of Bell, Magendie, and the decisive work of Joh. Müller and that of Panizza,**

---

* This is simply the Greek for 'common sensibility'

** B. Panizza (1785-1867), Professor of Anatomy at Pavia, best known for his work on the physiology of the brain.

have also shown that physical damage of the ventral roots of the spinal cord (containing the motor nerves) causes no pain: and, moreover, since surgical operations on man, or vivisection on mammals, have demonstrated that the touch-organs and their nerves, if damaged, evoke more pain than do most other parts: it appears then that the view expressed by Joh. Müller has much to its credit, when he says that we can feel pain only via the sensory nerves, and that touch is provided by sensory nerves which, by virtue of being supported in some parts of the body by specialized touch-organs in which they are very numerous, can also provide us with sensations of pressure, warmth, and cold, in addition to those of common sensibility. [121]

Commendable though these theories may be, they are still not wholly incontrovertible.

Magendie's experiments are, as he himself believes, not completely definitive, or at least there are still uncertainties to be cleared up concerning his experiments with birds. Moreover the fact that physical damage to the retina produces no pain cannot yet be taken to imply, with any certainty, that no other kind of stimulation will evoke pain in this area. On the contrary, the frequently occurring cases of pain in the eye, associated with photophobia, seem to prove that light can in fact evoke pain by way of the optic nerves. It could indeed be objected that such pains arise, not by way of the optic nerves, but by way of other nerves. These kinds of eye pains can, in fact arise from other types of nerves: but they should not be confused with those pains caused directly by light and evoked the moment light enters the eye. For since light acts directly only on the retina, and makes no impression on any other nerves, even if the latter be exposed, such pains can only arise with the involvement of the optic nerves. If we were to assume that in these cases too conduction of light impressions to the brain caused no pain directly, but that the impression itself could involve other nerve-vessels, causing pain by way of these, then we must note that, should this be the case, we should have to accredit the retina with common sensibility and with the ability to evoke pain. [122]

For there are many other cases in which pain is caused by the spreading, in the brain, of transmitted nervous impressions to other nerve-fibres. For one thing, photophobia and eye pains are sometimes observed where the damaged eye has been blinded by amaurosis to the point where light is not perceived as light and colour but only as pain. Ph. v. Walther[45]* has written on the subject:

---

* Ph. von Walther (1782-1849), doctor at Bonn and Munich, known for his studies of comparative physiology and of vision.

"Photophobia in many, even completely, amaurotic patients persists therefore in a most distressing manner and the impression of light remains extremely painful, even if they are no longer capable of distinguishing light from darkness." In such cases it seems to me that the retina is indeed sensitive to light impressions, and that the fibres of the optic nerves seem to conduct them to the brain. But in the brain there seems to be a hindrance to their being perceived as light and colour, and the light-impressions, if they are too intense, then can cause pain, without a concurrent sensation of light. The reason that specific sensations do not cause pain more often in the eye and ear is perhaps that light and sound impressions are not generally among the stronger types of stimuli acting upon our body, and possibly that mechanisms are provided whereby overly intense impressions are attenuated; for instance, by the contracting of the pupil, or by the muscles controlling the ear drum.

Experience also teaches us that a given part of the body may be extremely sensitive to one category of influences, yet be quite insensitive to another category. The muscles, for instance, as we shall demonstrate later, are parts of the body in which common sensibility is very acute; yet they are insensitive to physical damage or heat and cold.

The view that the touch-sense is only a more elaborate development of the nerves and organs contributing to common sensibility, and is therefore only a developed sense of feeling, might also be countered by the experimental observations that the use of ether and chloroform can reduce or even curtail the ability to feel pain, whereas the ability to feel contact persists: this unusual situation can also occur under other circumstances as in lead poisoning, for example. [123]

Gerdy[46]*, who has made observations on himself with regard to the effects of ether, reports that he first felt a sensation of warmth and numbness, as if alcoholic fumes had reached his brain. At first he noticed a numbness in his feet and toes, then in his legs and arms. In the most sensitive parts of his body the numbness was accompanied by a feeling of warmth and trembling. Only then came an insensibility to the feeling of pain. Gerdy took this to mean that smell, and taste, and sensations of touch, and tingling, were not curtailed by this general feeling of numbness, though pain definitely was.

Longet[47]** says, referring to the experiments of

---

* P.N. Gerdy (1787-1856), Professor of Pathology at Paris.

** F.A. Longet (1811-1871), Professor of Medicine in Paris, best known for his work on motor nerves and on the brain stem.

both Malgaigne* and Velpeau:**

"Some people remain conscious, and their external senses and even the sense of touch continue to function, while general sensibility (*sensibilité générale*) is the only thing curtailed. Others experience complete sensibility, but thinking is only slightly impaired, so that the patient understands questions directed to him although he cannot immediately answer them. He has lost his sensitivity to touch (*sensibilité tactile*), yet is still sufficiently conscious to be able to describe the experiments of which he is part, even being capable of sticking needles into his own flesh."

"Some patients can feel a tooth being touched by the dentist with an instrument, yet feel no pain when the tooth is extracted. Others can perceive the tearing of the tissue in the region of the parotid gland, caused by the instrument and yet they feel no pain." [124]

Pirogoff[48]*** says, basing his remarks on his own experiments:

"Etherized patients more or less remain conscious and retain their faculties, but they completely, or to a certain degree, lose their common sensibility (*sensibilité*). One patient, on whom I was performing an operation to produce a recto-vesicular fistula, talked during the whole operation, heard and saw everything, kept his thighs tightly together, yet only felt that the operation was being performed: he did not experience any pain."

If these observations are any indication, we must assume that the principal organs of touch lie in a different part of the brain from that of the principal organ of common sensibility: and therefore that the latter can be dulled by the action of ether, while the other is undisturbed in its function. It would then follow that the two capacities are to be distinguished.

The poisonous effects of lead also lead us to a similar conclusion. In the Hotel Dieu in Paris, J.H. Beau[49]**** observed a worker who had been occupied in a house painting and who had developed cachexia. The upper and inner part of his thigh was quite numb. If he were pinched or pricked, he felt neither contact nor pain: in other parts of the body he could feel contact, but still no pain. "Even the lightest contact did not escape him: his sense of touch was not lost." Since then Beau has found

* J.F. Malgaigne (1806-1865), Professor of Surgery in Paris.

** A.A.L.M. Velpeau (1795-1867), Professor of Surgery in Paris and author of books on embryology, obstetrics and other medical matters.

*** N.I. Pirogoff (1810-1881), Professor of Surgery at Dorpat and at Petersburg Academy, the first to introduce ether into Russia.

**** J.H. Beau (1806-1865), French doctor who worked at the Salpêtrière and various other hospitals. He specialized in heart and vascular disorders.

a more or less complete anaesthesia in at least 30 patients of this kind. He believes he can distinguish between two kinds of anaesthesia, anaesthesia for contact (touch-sense) and anaesthesia for pain (common sensibility). In the latter the patient felt no pain, no tickling of the type leading to sneezing, if the mucous membrane of the nose were stimulated with a soft feather, and no tickling of the type leading to retching if the uvula were stimulated; yet he could feel the contact. [125] There are various degrees of anaesthesia for common sensibility; according to Beau, the feeling of pain is either completely lost or is only more or less dulled. Anaesthesia for common sensibility is never absent when anaesthesia for touch exists, but it can exist without the latter. Anaesthesia for common sensibility often extends over a considerable area of the body, whereas anaesthesia for touch is always restricted to a small area of the body. Sometimes it is not possible to make pain artificially felt in a given region, yet great pain arising from the region itself can be experienced, e.g. in arthralgia. Beau, however, did not observe these two kinds of anaesthesia merely in cases of lead poisoning: he also found them in cases of hypochondria. For example, an 18 year old shoemaker, apparently completely healthy otherwise, felt no pain whatever in either arm when he was pinched, pricked, or touched with a white-hot iron: yet he was aware of the touch of a feather. Beau made similar observations on a 23 year old carpenter.

Some considerable time ago, the Geneva physician Vieusseux[50] observed and described a similar state in his own case of paralysis with complications. Common sensibility was lost on the whole of his right side, with the exception of the head: the touch-sense remained intact. On the head, it was not the right, but the left side which was affected. He could feel his pulse with his hand, but felt no pain if he was pricked or pinched with the finger nails. Spanish fly, and a serious abscess on the nail, accompanied by fever, only resulted in a feeling of warmth, tension, or itching. Warm water seemed cold to him, and cold water warm. The disease did not consist of a simple hemiplegia, for on the left side of the body sensibility was rather reduced, and the muscles were weak.[51]

Touching a limb which has 'gone to sleep' is known to cause feelings of prickling and quivering sufficiently unpleasant for contact with other bodies to be avoided, such as placing the foot on the floor. [126] Patients suffering from hemiplegia often find the rubbing on of ointments to be painful, although this would not give rise to the slightest feeling of unpleasantness to a healthy patient. In most cases the touch-sense is numbed, and it would therefore appear that common sensibility is increased in its sensitivity to the degree that the touch-sense has become insensitive. I have already tried to explain this striking phenomenon (p. [58]): I do not believe that we can conclude from this that different nerves are responsible for common sensibility and for the touch-sense in the skin.

Beau believes that a 'reflex' of sensation has to occur in order for pain or tingling to take place. If contact led to no sensation, then there could be no reflex for these sensations. The ability to feel pain must therefore always be absent if the ability to feel touch were absent. The ability to feel contact could, on the other hand, persist even though the ability to feel pain were to be curtailed, i.e. if circumstances existed under which the reflex for the sensation were prevented. Beau asserts that if one hits oneself with a stick on a corn, the pain occurs 1 or 2 seconds after the feeling of impact. If there were any substance to this observation, it would be highly favourable to the hypothesis offered by Beau. For the conduction of impressions through the sensory nerves occurs so rapidly that the time elapsing cannot be observed. Similarly, the conduction of impressions formed in the motor nerves to the muscles is, as my brother Eduard[52] observes, so rapid that the time between the stimulation and the resulting muscle-twitch cannot be perceived. On the other hand, however, a measurable interval does elapse, when an impression made on the sensory nerves travels to the motor nerves. [127] We have a good opportunity to observe the latter in a number of people simultaneously if we watch the women in the audience at a concert when, after a soft melody or a pause, the drums and trumpets suddently break in with full force. The visible reaction of the ladies, I have observed for many years, occurs measurably later than does the impact of the intense sound on the ear. In order to test Beau's theories, I have often hit my fingers hard, as they rested on a table, and found that sensations arising from the blow increase in strength, reaching their highest intensity noticeably later than the blow, after which they rapidly fade. But if the blow is delivered not to the finger-nail but onto the skin of the second finger-joint, I find that the pain is strongest immediately at the moment of impact, and therefore am led to assume that in the former case the pain arises somewhat later because of the protection afforded by the nail. As a result of the blow, the nail itself is dented or deformed. Perhaps the deformation itself does not directly cause the pain, so much as the movement whereby the nail, by virtue of its elasticity, resumes its original shape: as a consequence the pain is not felt at first. I do not mean to suggest by this that the sensation of contact and the sensation of pain arise simultaneously. If a person's back is stroked with a feather, he shudders; but this shudder is, according to my experiments, not simultaneous with the sensation of touch, but follows subsequently. The same holds for the tickling sensation before a sneeze. I believe that, just as with these other sensations, pain arises from a spreading of the impressions imparted to one part of the brain to other fibres of the brain, for I note that the area that seems to give the impression of pain also seems to be much more extensive than is actually the case, and this area, as long as no pain is involved, appears to be much more restricted.

VARIATIONS IN INTENSITY IN COMMON SENSIBILITY [128]

Only those parts of the body equipped with nerves have common sensibility. Areas lacking in nerves, such as the hair, the outer layers of tooth enamel, the epidermis, and the nails, are lacking in common sensibility both in the healthy and in the diseased state. Hair can burn without our feeling it; pieces can be filed from a tooth; and only if the enamel covering the interior of the tooth and the pulp is so thin that the tooth is no longer protected against heat, cold, and other influences, does the tooth become sensitive. Similarly, a fairly thick layer of nail-substance, or a fairly thick layer of superficial skin from the palm or bottom of the foot, can be shaved off without any sensations being felt save those easily transmitted to very sensitive areas *via* pressure and vibration. Nails can be knocked into horses' hooves to a certain depth.

If nerve-trunks supplying an area are completely severed, common sensibility in the area disappears. For example, if the nerves leading to the lowest joint of a horse's leg are severed, the hoof can be torn off with pliers without the horse showing any pain. This was observed by Professor Renner in Jena.

Common sensibility is most acute in the touch-organs and in the muscles of those body-areas most richly supplied with nerves. But the acuity of common sensibility is not shown in the same ways in these two cases. In the touch-organs pressure, and mechanical and chemical damage such as is caused by crushing or burning, evoke severe pain: but according to Bichat's* experiments, these stimuli do not give rise to pain in the muscles. Instead, pains are caused in muscles by too long a violent contraction increasing to the point where they are no longer bearable. Severe pains also result from cramp-like contraction of some muscles, as in calf-cramp or rheumatism. [129] Moreover, in many muscles there is a highly acute feeling of strain, whereby the muscles can be used, just as can

---

* M.F.X. Bichat (1771-1802), surgeon at Lyon and Paris, best known for his studies on animal tissues.

a sense-organ, to estimate resistance offered to our movement. It is by way of the exceptionally acute common sensibility in the vocal muscles that the exact degree of pressure needed in singing can be estimated: this is necessary if the vocal cords are to be tensed to the correct degree for producing a tone of a given pitch.

Between areas entirely lacking in common sensibility and areas possessing highly developed common sensibility are intermediate areas in which these sensations are so blunted as to be scarcely noticeable in the healthy state: they may however appear in certain pathological conditions.

COMMON SENSIBILITY IN THE SKIN AND OTHER SENSE-ORGANS

PAIN CAUSED BY HEAT AND COLD

In order to explain more fully the important theory on the nature of pain, it is necessary to investigate as accurately as possible the conditions under which heat, cold, pressure, and tension produce pain, and how pain arises as a consequence of persistent, uninterrupted muscular activity, even if moderate. In these cases, the transition from the tactile sensations of heat, cold, and pressure to that of pain can be observed and the steps whereby the sensation of strain in the muscles changes to fatigue, and fatigue to pain can also be observed, and the extent of the influences producing pain measured.

The pain experienced when one dips a hand into hot water should not be considered a touch sensation only differing from the touch sensation of moderate warmth in terms of its intensity. The pain arising from heat and cold is quite distinct from the sensation caused by heat and cold. Provided the former is not too intense, one can also experience the heat or cold giving rise to it and the pain originally caused by heat can be distinguished from that originally caused by cold. [130] If, however, the pain is intense, for example if tooth pulp is exposed to heat or cold, the same sensation is aroused whether the cause was heat or cold.

That the pain arising from heat or cold is based on a process different from that giving rise to sensations of heat or cold is indicated by the following situation. From the moment heat or cold cause pain, the ability to sense heat or cold in the affected region is, for some time, reduced or even completely abolished. I have already mentioned (p. [37]) that under a certain degree of heat or cold the power of the nerves to conduct is reduced or abolished. If a hand is removed from hot water sufficient to evoke a burning sensation, and is touched with a cold object or plunged very rapidly for a few seconds into cold water, the cold of the latter is not felt. If the hand is plunged rapidly several times into cold water, the ability to feel the cold gradually returns, so that on the third immersion a faint sensation of cold is evoked; this increases in intensity and distinctiveness on the fourth and fifth immersions.

The difference between the pain and the sensations of heat or cold here is partially due to the fact that the sensation of pain seems to extend over a larger area of the body, and that therefore the site of the sensation is not as clearly delineated as is the case with the site of the sensation of moderate warmth or cold. This is most clearly illustrated when pain is aroused by cold. If for example the entire hand be immersed in a large quantity of water of temperature 5°R (6.2°C) long enough for pain to be evoked, this pain is not restricted to the immersed region but extends up as far as the middle lower arm. This is an important phenomenon. I am not inclined to believe that the cold spreads directly up to the middle lower arm, causing pain there. Rather, I think that the cold, numbing the peripheral endings of the touch-nerves, penetrates to the nerve-trunks and, by way of their numerous fibres, makes an impression on the brain which is not limited to the brain area in which these fibres end, but spreads to neighbouring areas of the brain. [131] The sensation aroused in these neighbouring areas is then transferred in our imagination to the lower arm. As has been pointed out on p. [23], there are other cases where pain is not restricted merely to the area affected by the pain-evoking stimulus.

Pain, therefore, appears to be evoked by heat or cold when the effective source of this pain gives rise to such a strong impression on the brain that the impression spreads more extensively over the brain. Whether the impression on the brain will reach such an intensity depends on five conditions: 1. *the degree of the heat or cold affecting it:* the higher or lower the temperature, the more rapid and intense is the onset of pain; 2. *the duration for which we are exposed to the heat or cold:* the longer we are exposed thereto, the more our organs take on the high or low temperature, and the more this temperature will then penetrate to the interior of those organs, affecting not merely the peripheral nerve endings, but also the nerve-trunks and, *via* the multitude of their fibres, causing similar impressions to be formed in the brain, where they summate to give an intense impression; 3. *the sensitivity of the body region* exposed to the heat or cold: for example, we feel stronger and more immediate pain if we dip the tip of the tongue into hot water than we do if we dip the finger into the same hot water; 4. *the surface-area of the sensitive region* exposed to the hot or cold influence: the greater this area, the greater the number of nerve-fibres which simultaneously receive the impression of heat or cold, with these many impressions summating in the brain to a single strong impression which can be so intense as to spread and there evoke pain; and finally, 5. *the thinness or thickness of the protective layer of epidermis.* This is a poor conductor of heat, but is known to vary in thickness over the various portions of the skin; for the thinner the epidermis, the more rapidly heat or cold can penetrate to sensitive areas.

As for the *degree of heat or cold necessary to evoke sensations of common sensibility or even pain*, it would appear that this would be that which, having affected the nerves for some time, limits their conduction capacity, or even abolishes it, and thereby even evokes sensations of common sensibility. [132] If intense enough, these might increase to a painful level.

A temperature of 39°R, if it has acted long enough on our organs, reduces their conductive capacity. It is approximately this same temperature which can evoke sensations of common sensibility and moderate pain. My experiments have not been extensive enough to enable me to determine the degree of cold necessary to reduce markedly the conductive capacity of the nerves, but I can say this much, that a temperature of 9° or 10°R can have this effect provided the water be given enough time to exert its effect, and the immersed area be large.

If we dip one hand into moderately hot water, the sensation upon immersion is at first vivid, but immediately fades; but then it gradually increases to a level so painful that we are obliged to withdraw the hand. The lower the heat, the longer it takes before pain is felt. As the water temperature cannot fall below 0°, we have enough time to observe how the sensation, here too a vivid one at first, decreases, but then increases again, eventually to the level of painfulness: but it always takes longer in this case before pain is felt, and it never reaches an intolerable intensity. Other sensations of common sensibility precede the pain in both cases. In a hand dipped in warm water of about 40.5° or 41°R, a tension or swelling is felt: and a discomfort is evoked which seems to arise from the stronger pulsations of the small arteries and from the twitching of single bundles of muscles. The same twitching can be observed in the tip of the tongue, if it be immersed: severe cold can also evoke it.

When I held the middle finger of one hand in water at a temperature of 40.5°R for 2½ minutes, whilst I held the same finger of the other hand in water of 9°R, the sensation of cold was at first stronger than that of warmth: but after a few seconds, both sensations were equally strong. [135] Following this, however, the sensation of warmth increased until pain began, with a throbbing being felt in the finger: and when I then put both fingers together and felt each with the other, I could feel neither the warmth of the warm finger with the cold one, nor the cold of the cooled finger with the warm one: *for the sensitivity of both fingers had been reduced* so that I felt neither warmth nor cold. If the temperatures of the two fluids were +30.5^{o} and +9°R, and I put the two fingers together after 10 minutes immersion, cold was the sensation which I was able to feel.

The cooled finger was reduced in sensitivity and provided no sensation of warmth from the other finger it touched: the warmed finger, on the other hand, maintained its sensitivity, and I could feel the cold of the cooled finger with it. If the two temperatures were +41°R and +19°R, with the right hand exposed to the one and the left hand to the other, the water, after a short while, evoked a burning sensation which after 28 seconds became one of pain. If both hands were now brought into contact, the warm temperature of the heated hand could be felt by the moderately cooled hand, but the coldness of the cooled hand could not be sensed by the heated hand: for the sensitivity and conductive capacity of the nerves of the latter had been reduced or abolished.

Water exceeding the blood temperature by about 20°R causes, after only 4 to 6 seconds, a violent and intolerable pain in the joint of a finger immersed therein. On the other hand, water 20°R colder than the blood evokes, after long contact, only a feeling of numbness, with no real degree of pain. Even a mixture of snow and water, at a temperature of 0°R (some 30° colder than the temperature of human blood) only evokes moderate pain after a long time, and which can easily be endured. *Heat therefore evokes pain more rapidly than does cold, and the pain is more intense.*

That the area of skin surface exposed to a hot or cold temperature influences the pain evoked can be seen from the following experiments. [134] If the joint of a finger was held for a fairly long time in water at a temperature of 39°R (48.7^{o}C), I felt no marked pain, but had only sensations of common sensibility which could be endured without discomfort: but if I put my whole hand in, I felt pain. Similarly I found that when I dipped a finger-joint in water with a temperature of +5°R (6.2^{o}C), I felt no pain, but I did when I immersed the whole hand.

In order to determine more accurately the effects of the time necessary for pain to be elicited, that is, the time necessary for the heat or cold to penetrate more deeply into the body and stimulate the nerve-trunks, I had an observer dip the last joint of the index finger into hot water and counted the seconds elapsing before the pain reached a level necessitating his withdrawing his finger from the water. If the water was not very hot, he felt no pain for some time: quite a few seconds had to pass before he felt any. The results showed that the finger-joint could be exposed for a longer time to hot water the lower the temperature of the water.*

Temperature	No. of seconds before pain necessitated finger-withdrawal from the water	Temperature	No. of seconds before pain necessitated finger-withdrawal from the water
$45\frac{3}{4}$° R	11	52° R	3
$45\frac{1}{2}$° R	$13\frac{1}{2}$	51° R	4
$44\frac{1}{2}$° R	14	$50\frac{1}{3}$° R	$4\frac{1}{2}$
$44\frac{1}{2}$° R	14	$49\frac{2}{3}$° R	$5\frac{1}{2}$
44° R	21	49° R	$5\frac{1}{4}$
$43\frac{1}{2}$° R	20	48° R	7
$42\frac{1}{2}$° R	23	$47\frac{2}{3}$° R	7
$41\frac{1}{2}$° R	no withdrawal	47° R	9

When I carried out such experiments on myself, I found that the finger, with repeated immersions in hot water, becomes increasingly insensitive to the effects of the temperature, and that it was therefore better if different fingers were used, immersing the end joint of the various fingers.** [135]

Temperature	No. of seconds elapsing before the pain became too intense	Temperature	No. of seconds elapsing before the pain became too intense
57° R	$3\frac{1}{2}$	56° R	$2\frac{1}{2}$
53 "	$4\frac{1}{2}$	55 "	$3\frac{1}{2}$
52 "	4	54 "	$3\frac{1}{2}$
51 "	5	53 "	4
51 "	4	52 "	4
50 "	4	51 "	5
49 "	8	50 "	5
48 "	$5\frac{1}{2}$		

Fifteen years later I again made the same observations on myself, without checking the earlier results beforehand.

---

* The table below is almost the same as the one on p. [127-8] of *De Tactu*.

** The table below is identical to the one on p. [128-9] of *De Tactu*.

Temperature	No. of seconds elapsing before the pain became too intense	Temperature	No. of seconds elapsing before the pain became too intense
70°R	$1\frac{1}{2}$ to 2	53°R	7
68 "	3	52 "	7
66 "	3	51 "	8
65 "	3	50 "	9
63 "	3	49 "	10
62 "	3	48 "	12
60 "	4 to 5	47 "	14
59 "	4 to 5	46 "	17
58 "	5	45 "	23
55 "	6	44 "	28
54 "	6		

In experiments in which another observer used the last finger-joint it emerged that the onset of pain occurred rather earlier if a finger on the left hand was immersed than was the case with a finger on the right hand, probably because the epidermis on the right hand, used for harder types of work, was thicker than on the left.

Finger-joint on the right hand		Finger-joint on the left hand [136]	
Temperature	Seconds	Temperature	Seconds
59°R	4 to 5	59°R	4
55 "	6	55 "	5
52 "	6	52 "	5

If the tip of the tongue was dipped into hot water, the onset of pain necessitating my withdrawing the tongue was more rapid than was the case for the finger.

Temperature	Seconds	Temperature	Seconds
50°R	2	42°R	18
48 "	4	42 "	17
47 "	4	41 "	19
46 "	6	43 "	12
45 "	7	42 "	18
44 "	8		

These experiments illustrate 1. that, in order for pain to be felt a limb had to be exposed for a longer duration to water of a hot temperature, the lower the temperature was, 2. that in water which was only mildly hot, necessitating 10 to 28 seconds before a sensation of pain was elicited, a change of only 1°R or $\frac{1}{2}$°R in the temperature caused a marked difference in the time needed before pain was felt, whereas in very hot water, 1°R change made little difference in this respect; 3. that the limb was made less sensitive to hot temperatures if it had been previously exposed to hot temperatures.

In order to investigate the pain caused by bodies whose temperature fell below 0°, I carried out the following experiment. I dipped a very large iron key, the solid shaft of which had a rounded end and was of diameter 4 Paris *lines*, into mercury of -4.2°R to -2.1°R until the key's temperature reached that of the mercury. As the very large bit of the key was situated near this end, it maintained its very low temperature for some time when brought into contact with the skin. In many parts of the face, for example below the zygomatic bone, and at the corner of the mouth, a stinging pain resulted, similar to that of burning, in many other places the sensation was neither stinging nor burning, but was different from that of cold. [137] Nowhere was it so intense as to be intolerable. In many places on the head and face, and in very many places on the rest of the body, there was no pain felt at all. The pain seemed to arise either where the skin was very sensitive, with the epidermis also being very thin, or at places where the nervous trunk containing sensory nerves lay just beneath the surface of the skin. Quite remarkable were the marked differences between many skin-areas lying very close to each other, often one area would give rise to pain when contacted, while a neighbouring area would give rise to none. It might be surmised that these differences arose if the cold affected the nerve-trunks of sensory nerves lying beneath the skin in certain individual areas. Moreover, unknown organs whose function may be to provide sensations of warmth do not seem to be evenly distributed in the skin.

Experiment Number	Temperature of Key	Body-area touched by the key	Whether pain was felt or not
1	-4.2°R	Forehead, mid-line above nasal proturberance (Glabella)	A strong feeling of cold, but no pronounced feel- of spreading pain.
2	-3.2°R	Forehead, 5.3 Paris *lines* from the mid-line, just above the Arcus supraciliaris, right side.	No pain even though the round-end of the key was kept in contact for one minute.

Experiment Number	Temperature of Key	Body-area touched by the key	Whether pain was felt or not
3	$-3.2^{o}R$	Forehead, 6 Paris *lines* from the midline, 7 Paris *lines* above the middle of the Margo supraorbitalis, right side	Mild pain. I could feel the artery pulsating
4	$-3.2^{o}R$	Forehead, 11 Paris *lines* from the midline, right side	After 10 seconds pain was felt, which became very noticeable after 20 seconds. I could feel the artery pulsating
5	$-4.0^{o}R$	Forehead, 11 Paris *lines* to the left of the midline, 6.5 Paris *lines* above the Margo supraorbitalis.	Pain was felt after only 8 seconds extending into the eye-socket
6 [138]	$-3.7^{o}R$	Right forehead, at the region of the branches of the supra-orbital nerve and supratrochlear nerve.	Pain after 14 seconds
7	$-4.0^{o}R$	Forehead, above the outer part of the orbit, 2 inches 2 Paris *lines* from the midline, and 5.5 Paris *lines* higher than the uppermost edge of the Margo supraorbitalis	Fairly severe pain

Experiment Number	Temperature of key	Body-area touched by the key	Whether pain was felt or not
8	$-3.7^{o}R$	Upper eye-lid	Pain after 8 seconds, more vibrant and localized than was the case with the conjunctiva covering the eyeball
9	$-2.9^{o}R$	Face below the zygomatic bone	Stinging pain, similar to that of burning
10	$-2.1^{o}R$	Corner of the mouth	After a few seconds, stinging pain similar to that of burning
11	$-2.6^{o}R$	Ear-lobe	No pain
12	$-2.6^{o}R$	Tip of tongue	Pain similar to that caused by burning the tongue
13	$-4.0^{o}R$	Midpoint of the left incisor of the upper jawbone (this was quite healthy).	Mild pain after 2 seconds, not increasing; the sensation of cold persisted some time after removal of the key
14	$-3.9^{o}R$	Hard palate behind the incisor	No pain
15	$-3.9^{o}R$	Posterior hard palate in the region of the spina nasalis posterior	No pain
16	$-3.2^{o}R$	Posterior hard palate anterior to the hamulus pterygoideus	Sudden onset of pain (probably because of the proximity of the branches of the pterygo-palatine nerve)

Experiment Number	Temperature of key	Body-area touched by the key	Whether pain was felt or not
17 [139]	-2.6°R	Arm, between condylus internus and olecranon, where the ulnar nerve lies close below the skin-surface	At first only a mild feeling of cold: but after 11 seconds the sensation increased and gradually became painful, reaching a severe level after 40 to 50 seconds. The pain also involved part of the lower arm
18	-2.4°R	Outer part of the olecranon	Pain was also felt here
19	-2.6°R	Lower arm, just about the middle of extensor digitorum communis	Moderate pain
20	-2.4°R	Knee-cap	No pain
21	-3.7°R	Right side, below the tendon of the biceps, when the peroneal nerve lies close to the skin-surface; the key was pressed along the full length of its shaft	Pain, which seemed to include part of the calf, but not the foot

PAIN IN THE SKIN CAUSED BY PRESSURE AND TENSION

Experiments similar to those carried out on the elicitation of pain by temperature-changes may also be carried out on the elicitation of pain by pressure and tension. For example such experiments can determine the degree of pressure which must be exerted on a square inch of skin supported by bone in order for the sensation of pressure to change into one of pain. One can also find out experimentally the effects of hanging a weight on 4, 8, or 16 hairs, and how heavy the weight must be in order for the tension to cause pain in these various cases.

Pain arising from pressure, e.g. from tight shoes and boots pressing against corns, are amongst the most severe and persistent occurring in the skin. [140] Pus, accumulated without exit under hard skin, also causes severe pain, which ceases immediately when the pus finds a way out.

Less instructive are observations on the pain caused by cutting the skin with a sharp knife, by squeezing it, by the harmful effects of caustic potash and similar corrosive solutions, as well as a red-hot iron. These all cause physical or chemical damage to the skin and its nerves.

The act of cutting causes pain, not the mere fact of having been cut (before any resulting inflammation occurs, which itself can cause pain because of the changes it brings about). If the damage occurs very rapidly, as with a bullet wound, the pain can be so slight that the person wounded does not even realize that he has been hurt. Many situations giving rise to pain cause, after a short time, the nerves involved to lose their sensitivity, e.g. to heat, cold or pressure. Persistent pain may nevertheless result from such situations insofar as the object causing the pain penetrates increasingly deeply into the body, thereby stimulating new nervous regions: or the effects of the object and the pain it causes may ease off from time to time, during which time the nerves recover their capacity to conduct.

PAIN FROM ELECTRICITY

Because a galvanic current, if the circuit be open and closed, affects not only the motor nerves, and can most effectively cause the muscles to contract, but also affects sensory nerves such as the optic nerves, one would expect that the simultaneous effect of the current on the sensory nerves would also give rise to severe pain. While indeed the sensations evoked by electricity and a galvanic current are unpleasant, they are nevertheless much less painful than might have been expected if one compares the violent cramp they can cause in muscles, e.g. if a persistent tetanus in the flexor muscles prevents our opening the hand. [141] We have seen on p. [25] that if the elbow be dipped into a mixture of water and ice, the cold can penetrate through to the trunk of the ulnar nerve. As this nerve contains numerous sensory as well as motor nerves, and as the cold reaches both to an equal degree, it is interesting that the effect of the cold on the motor nerves is hardly noticeable, whereas there is quite a severe pain as well as numbness. Heat and cold therefore appear to cause pain more readily, while electric currents cause muscle-contractions more easily.

SHUDDERING AND TICKLING IN THE SKIN

If several areas of the skin, e.g. the skin of the back, are lightly stroked with a feather, not only is contact felt, but, after the sensation of contact ends, there is a 'shudder', i.e. a sensation *lasting for some time*. This sensation seems to be restricted not merely to the touched area, but extends to other areas which were not touched; moreover, when the sensation has subsided in one area, it may be felt again in a neighbouring area and this seems to be the manner in which it travels. We say then that 'a shudder runs over us'*. As the peripheral endings of the nerves do not intercommunicate in such a way that the impression might be transferred from one nerve to another, it may be surmised that it is in the brain that the impressions are transferred to neighbouring areas, and that the mind merely imagines that these latter sensations arise in the areas of skin adjacent to the areas originally stimulated. It is noteworthy in this context that a shudder does not arise at the moment of contact, but requires an observable amount of time before it occurs and spreads. With other situations, however, a simple transferral of sensations occurs so rapidly that no interval of time appears to elapse. This phenomenon is similar to that in reflexes, when the impressions from the sensory nerves require a perceptible time before being transferred to the motor nerves. [142] If the thumb-nail be used to stroke the middle of the backbone, with some considerable pressure being exerted, a shudder is evoked even when the subject is fully dressed. At the same time, the dorsal muscles which keep the spine upright occasionally contract. A similar shudder arises if some other areas of the skin are touched with a cold object. The shudder here is often called a 'chill', but it is not to be inferred from this that a spreading of the cold in the skin is the major cause of the spreading sensation, which persists even after the contact has been broken. Without other previous stimulation of the skin, a chill can accompany inflammations and fevers, often in the shivering stage of a cold fever, a stage during which some muscles often contract as well. Here other influences appear to excite activity in the central nervous system, interpreted by the mind as a sensation in the skin.

A light touch with the tip of a feather on the lips, the edge of the nostrils, and the regions about the face elicits a characteristic sensation of tickling. This

---

* German, *es überlaufe uns ein Schauder*.

sensation also persists after the period of contact, but in addition will occasionally increase in intensity, and fluctuate from place to place, causing an instinctive desire to scratch or rub the skin. Something similar can be observed *if the mucous membrane of the nose is lightly stimulated*. This kind of tickle increases to the point where we are forced to sneeze because the stimulation causes a contraction of the ducts of the mucous glands, forcing their contents outward, so that the tickling sensation aroused by these contents in turn renews the stimulation. Something similar possibly happens in the skin where the ducts of the skin glands are concerned. *The ear passage is not susceptible to tickling*, but is very sensitive to the touch of cold objects: this latter is not the case with the mucous membrane of the nose. *The intense sensation excited by objects coming into contact with the inner surface of the eye-lid* also seems to be a particularly severe kind of tickle, but one which easily reaches a painful level, eliciting a flow of tears. The part of the conjunctiva covering the sclerotic coat and the cornea is not very sensitive or susceptible to tickling. [143] It can be touched without pain or tickling being felt. The eyes can be opened under water without any unpleasant sensations arising, and even salt-water evokes only a mild sense of stimulation in the inner corner of the eye, at the punctis lacrymalibus. On the other hand sulphuric acid or ammonia fumes cause an intense sensation, but one which cannot be called tickling. The numberous hairs of the skin are instrumental in transmitting slight impressions to the interior of the skin and to the glands, yet their presence is not essential for the evocation of a tickling sensation; for tickling can be experienced in many regions of the body, e.g. the bottom of the foot, the palm, and the palate.

Why only certain regions of the skin and mucous membranes are ticklish while many other regions lying close by should not be so, or only to a slight degree, is difficult to explain. For my own person, the lips, skin and mucous membrane of the nasal entrances, the skin of the hard palate close behind the upper incisors, the skin of the bottom of the feet, and the skin in the arm pit are particularly sensitive areas. Areas with a finely developed sense of tickle are not always provided with a well developed touch-sense, for example, the mucous membrane at the entrance to the nostrils.

SENSATIONS OF COMMON SENSIBILITY IN THE SKIN ARISING FROM VASCULAR MOVEMENTS, SECRETIONS OF JUICES FROM THE BLOOD, AND THE PROCESSES OF NUTRITION

It is self-evident that sensations caused by the flow of blood and its concomitant processes must belong to the class of sensations of common sensibility, and is not that of sensations involved in the perception of objects. For, if the objects exciting a sensation simultaneously affect even the smallest amount of the tissue forming the skin, all justification for the mind's interpreting the resulting sensation as originating from a body discrete from that of the sensation-organ disappears. In such circumstances we can only perceive our own changed state of sensitivity.

A mustard-poultice applied to the skin increases the amount of blood in the stimulated region of skin, and sensations of warmth and burning simultaneously arise. [144]

Skin-rashes can cause sensations of itching, burning, and countless other modifications of common sensibility. The fact that extensive rashes do not cause severe pain is to be ascribed to the fact that the damage to the skin caused by the rash occurs gradually.

The pain produced by nettle-stings, and the stings caused by bees, wasps, and other insects is primarily due to irritating substances entering the wound, from where they spread. Otherwise the intensity and persistence of the pain would hardly make sense, given the small size of the lesion.

COMMON SENSIBILITY IN THE MUSCLES

Bichat[53] writes: "In animals, it is usually the case that sensitivity is very poorly developed in the muscles. Amputations and cross-sections performed in experiments with living animals cause no particularly unpleasant sensations. It is only when a nerve-fibre is cut that marked pain results: the network of muscles is in itself only slightly sensitive. Various types of stimulation also cause no particular pain. Nevertheless one characteristic kind of sensation, that of fatigue caused by repeated contractions, occurs in the muscles". It is most instructive for students of the nature of pain to note that many stimuli which cause severe pain in other areas are ineffective when applied to the muscles, while other phenomena demonstrate a high degree of sensitivity in the muscles.

Amongst these are the pain resulting from severe fatigue, the extreme pain associated with tonic spasms, e.g. in the calf, and rheumatic pain, with its sudden onset and recurrence the moment the affected muscle is contracted. [145] But foremost of these is the subtle sensation accompanying many contractions of the voluntary muscles, in which we can feel the effort made during use of the muscles, and can accurately estimate the resistance offered during this activity.

*Even the movements of many organic muscles are associated with painful sensations, e.g. those of the uterus, called labour 'pains' or 'dolours', those of the rectum during tenesmus**, and those of the intestines during gripes. I think it probable that the violent contractions of the muscles of the empty stomach, reducing its volume, are associated with sensations forming one aspect of the general sensation we call 'hunger'. Feelings such as those of *nausea during vomiting*, the *desire to pass stools*, even the feeling characteristic of *diarrhoea* should also be considered as sensations accompanying large peristaltic and antiperistaltic movements. A number of pains associated with the heart appear to be caused by its movements related to sensation, and it seems highly likely to me that the *pleasurable feelings during the ejaculation of sperm* are associated with movements undergone by those muscular tissues of the seminal vesicles and prostate pointed out by me elsewhere. The muscles, closest to the sense-organs and particularly the skin, are amongst the parts of the body most richly supplied with nerves. They also possess some of the most acutely developed sensations of common sensibility, though only for certain types of stimuli.

In order to see how feelings of strain merge into those of fatigue, and those of fatigue into those of pain, one may see how long the upper and lower arm may be held out horizontally, not fully extended. This duration will be found to differ widely between different individuals, depending on their muscular strength and degree of extension of the arm, and also on the angle by which the arm approaches or diverges from the horizontal position. In my own case pain began after about 300 seconds, this increasing in severity so gradually, however, that I could still hold my arm up after 600 seconds, and indeed up to 900 seconds (a quarter of an hour). [146] I think I could have continued to hold it up for even longer. But I finally began to notice a slight trembling of the muscles. If, at this high level of fatigue, I repeatedly lifted my arm even higher, the pain did not increase during contraction of the fatigued muscles, but it did so later. While this strain was going on, I imagined that a relaxed position would be very pleasant, but was mistaken; for while indeed I was relieved not to have to continue exerting effort, I nevertheless found that the pain continued at the same level of intensity, and did not fully

---

* Abnormal inclination to void, but with little or no discharge.

subside even after two hours. It was even present if I lay down and allowed the entire limb to be as relaxed as possible. In this experiment there was no pressure of any extent exerted on the muscles or nerves, nor any over-extension of the muscles. The pain was caused by a compositional change in the nerves and muscles induced by an inadequate nutritive supply dependent only upon the length of time the muscle had been contracted. Pain can also be caused in other instances by a compositional change in the constitution of other areas of the body, for muscular contraction appears to be accompanied by a chemical process, much as if a gun had been fired. According to Berzelius and Liebig[54], lactic acid is formed in the muscles, either in a free state or as an acidic lactic salt: while the experiments of Bunzen[55], Becquerel and Brechet on warm-blooded animals, and those of Helmholtz[56]* on cold-blooded animals, indicate that heat is also generated. The lactic acid is perhaps a breakdown product.

The pain only seems to be located in those muscles utilized during the lifting of the upper and lower arm. For instance, the finger muscles were not involved, and could be moved afterwards without arousing any pain. [147] Lifting the upper or lower arm, on the other hand, gave rise to pain even after the limb had been relaxed. The *onset* of pain as a result of moderate, but uninterrupted and protracted activity of the muscles, despite there being any other source possible inducing pain, and the *long persistence* of the pain following relaxation of the muscles, argue strongly that the cause of this pain is to be found in the constitutional changes undergone by the muscles during such sustained activity, and which could only disappear gradually as the muscles are re-nourished. It is known that the considerable change in the muscles of game animals which have been run to ground, resulting from the intense and protracted strain, even show up in the appearance and taste of the meat itself.

Sensations of heat and cold, pressure and stretching, can, as we have seen, change into those of pain if they exceed a certain intensity. Similarly, sensations of effort in the voluntary muscles can change into those of fatigue, and these in turn into pain. This pain, even if located in quite different areas, and arisen from quite

---

* J.J. Berzelius (1779-1848), J. von Liebig (1803-1873) and A.C. Becquerel (1788-1878) were eminent chemists and physicists of Weber's era. H.L.F. von Helmholtz (1821-1894) later became known for his work on vision and hearing. T. Bunzen (1776-1807) was a Danish doctor known for his work on the effects of electricity on animal tissues. We have been unable to find information on Brechet.

separate sources, is nevertheless very similar to the pain resulting from over-pressure or stretching in some parts of the body. There is nothing specific about it. Compositional changes in the muscles and muscle nerves, apparently brought about by long-sustained exertion, and which, as noted, I take to be the source of pain, can also be induced by other means, e.g. fever. The similar type of pain induced by these other means is often called a weakness of the limbs.

The common sensibility in the muscles through which we feel the degree of effort exerted (necessary if we are to overcome any resistance which might be offered) is so subtle that it functions as a sense, which we might call the sense of 'force'. Above on p. [96] it was shown experimentally that we can distinguish between two weights more accurately and clearly through the common sensibility of the muscles than *via* the touch-sense. In the former case we can distinguish the heavier from the lighter of two weights in the ratio 40:39. [148] Now it could indeed be proposed that the source of the sensation of effort resides not so much in the muscle-nerves as in the part of the brain affected by the will. In this respect it is to be noted however that weight-differences can be perceived merely via the extension undergone by the muscles, and hence without any effort of the will. I tied a loop of cloth about the wrist and let my arm hang down over the back of a chair, so that it was supported under the armpit. If weights were now hung on hooks fixed to the cloth, the arm being held loosely, the stretching of the muscle, gave rise to sensations sufficient to permit of weight-differences being perceived. Admittedly, sensations of pressure also contributed to this, these sensations being located at the point where the hand was pinched by the cloth loop and where the arm was supported by the chair. The stretching of the muscles was due to the fact that the relaxed ligaments in the joints of the hand and overarm not involved yielded somewhat, so that the points of application of several muscles moved apart. It has already been explained that we come to learn by experience how much effort has to be exerted in certain muscles to place our limbs in certain positions and hold them there, and that this knowledge is so associated with the feeling of effort that, at any moment, we can obtain from this feeling an image of the position of our limbs, even when we cannot see them and they do not touch each other. If somebody places our hands or fingers in a given position, and this position is maintained, we can easily tell what kind of position they have assumed. From the feelings of effort in the muscles we can therefore tell at any moment the position of parts of the body intentionally moved, and also whether there is anything unnatural about a given position.

These feelings, for example, also enable us to tell

with great accuracy, the direction in which hair might be pulled (see p. [90]). The effort of certain muscles, produced when we sing a note or pronounce an articulated sound is associated with an imaginary concept of these tones and sounds, and there is no more vivid way of evoking such an imaginary image than to bring the vocal and speech organs into positions from which these tones and sounds would result if we were to exhale. [149] The effort of certain muscles used to make certain facial expressions is associated with the emotion which compels us to make such gestures. The reverse is also true, where a certain facial expression can evoke a certain emotion; this disappears most easily if the expression of the face is changed, e.g. if we smooth out certain wrinkles in the forehead with the hand.

Very fine branches of a sensory nerve, the ramus ophthalmicus of the trigeminal, have been traced to the eye muscles, which are known to be supplied by motor nerves, namely the oculomotor, trochlear, and abducens nerves. A few sensory nerves no doubt also lead to other muscles in addition to motor nerves: the former mingle with the latter by way of anastomoses of various nerves. Perhaps these sensory nerves are to some extent the cause of the vivid and acute common sensibility of the muscles. The fact that many voluntary muscles, such as the diaphragm, have very poor common sensibility might be explained as the result of their being supplied with only a smaller number of such sensory nerves. However, the assumption that common sensibility is always mediated by nerves originating in the touch-nerves appears to be unconfirmed if the kind of disability known as anaesthesia is considered. In anaesthesia, sensibility in the limbs is lost, yet the ability to move them voluntarily remains intact. But not every disease designated as anaesthesia is to be taken as such: for many muscles affect the limbs from a distance, and muscles affecting voluntary movements of the fingers, for example, may lie not in the fingers but for the most part in the lower arm, so that if the disability attacks only the hand, and the lower arm as well the fingers may still be moveable even though deprived of sensations. This is not true anaesthesia. But even in true anaesthesia, where a large area of the body is wholly or nearly wholly deprived of the touch-sense and the ability to feel pain, and yet can still be voluntarily moved, the ability to walk persists; this presupposes that a sense of equilibrium, without which balance becomes impossible, is still intact.[57] [150]

PARTICULAR SENSATIONS OF COMMON SENSIBILITY IN PARTS OF THE NERVOUS SYSTEM [151]

All sensations are based on changes occurring in the nerve-fibres, and to this extent it may be asserted that there is only one common sensibility in the parts of the nervous system. Nevertheless there are many cases in which it is possible to determine whether the first impulse causing such changes originated from something affecting the peripheral endings of the nerve-fibres, or the nerve-trunks or the central areas of the nervous system. It is important in a medical context to be able to make this distinction, but often very difficult.

Tremors, crawling sensations, numbness or insensitivity of the limbs, and pain caused by spasms have been considered as special instances of common sensibility in the nervous system, and indeed these symptoms often indicate an ailment which does not reside in the nerve-endings concealed in the skin or muscles, but is located in the nerve-trunks or in the brain and spinal cord.

Many pains certainly arise because this pain-producing source affects not so much the nerve-endings in certains areas as the nerve-trunks or the nerve-fibres in the brain and spinal cord. But the latter are difficult to distinguish from these other pains. I am not sure that headaches do not stem from brain-fibres, but rather from the endings of sensory nerves extending into the brain, whose function is to give the brain sensitivity. The existence of such *Nervi nervorum* has not yet been demonstrated. Longet, who once was involved in a controversy[58] with Magendie as to which of them had shown by physiological experiments that sensory nerves projected from the dorsal roots of the spinal nerves in the spinal ganglia over to the ventral roots, extending therein in the direction of the cord and thus making these roots sensitive, has now conceded the honor entirely to M. Magendie and has now abandoned his own experiments.[59] Formerly he had thought he had discovered that the ventral root of the nerves in the spinal cord was sensitive, so long as the dorsal roots were not severed; if the latter happened, the ventral root was rendered insensitive. [152]

COMMON SENSIBILITY IN ORGANS WITH MUCOUS MEMBRANES

As I noted on p. [146], some feelings of common sensibility in these organs are probably associated with pronounced movements of their muscle fibres; others are no doubt based upon impressions made on the mucous membrane itself.

Here, it is appropriate to mention the *tickling* sensations felt in the larynx, produced by even the least irritating fluids, and by all solid objects. This tickling results in the shutting of the glottis and in coughing.

In the case of bronchotomy, this high degree of irritability only exists in the larynx, and not in the trachea. The tickling sensation causing nausea, induced by touching the back of the tongue, may also be included among these sensations.

Solid bodies, as mentioned earlier, will evoke a peculiar tickle causing sneezing, if brought into contact with the *mucous membrane of the nose*. Water, on the other hand, does not irritate the mucous membrane, even if the nose be filled. Carbonated drinks do irritate the membrane however, if the air from them rises from the stomach and enters the nose. This brings on a very strong sensation in the nose, but not in the mouth. Sugar soaked in peppermint oil causes a peculiar sensation on the tongue, palate and gullet.

In the *ureter*, strong sensations are felt if blood or gritty deposits are present in the urine.

Thirst is probably based on a change in secretions present in some mucous membranes, arising in turn from insufficient water-content in the blood. When a large amount of salt is absorbed into the blood, thirst will therefore result, for the usual quantity of water in the blood will be inadequate to produce the necessary dilution of the blood. On the other hand, cessation of drinking will produce thirst more rapidly, the more water is lost through perspiration or excretion. The reason that we feel dryness predominately in the throat and palate is probably that these regions are particularly sensitive to certain influences, this sensitivity being absent in the oesophagus, intestines and even the tongue. [153] Just as the conjunctiva is particularly affected by sulphuric acid fumes and ammonia fumes, my experiments have shown that the same is true of the palate and the pharynx, though not to quite the same extent (see p. [119]). These areas are supplied with a touch-sense and also have a much thinner epidermis than has the tongue. If the conjunctiva were not moistened with tears, we would possibly also feel dryness of the eyes during thirst. While we feel thirst most in the palate and throat, the conditions causing these sensations are certainly not simply restricted to these areas of mucous membrane.

The *feeling* of a desire *to breathe*, following *a protracted exposure to poor air*, is undoubtedly not located merely in the mucous membrane of the lung, but also in the heart, for this lack of air is bound to be associated with disturbances in the circulation of the blood.

COMMON SENSIBILITY IN AREAS WHERE NERVES AND BLOOD VESSELS ARE NOT ABUNDANT

When healthy, cartilage, bone, serous tissue, and probably also synovial tissues, blood vessels, tendons, connective tissue, and fatty tissue, do not have very dense capillary networks, nor are these well filled with blood. In cartilage, if it is not ossified and no marrow-cavities develop, the blood-vessels cannot be made visible at all, and it is equally difficult to trace any nerves to cartilaginous tissue. On the other hand nerves have been anatomically mapped in sinus and synovial tissue, and in many tendonous areas such as the dura mater, the external walls of the arteries and in the walls of the large venous trunks. The intermediate and inner walls of the arteries do not, according to my investigations, possess any capillaries, but the longitudinal walls of the larger veins do.

All these regions appear capable of providing sensations of common sensibility only in the diseased state. Doubtless the blood vessels to which the nerves of these areas run are mainly responsible: over-stretching of the vessels resulting from an excess of blood seems to be a condition responsible for the pain felt. [154]

In order to investigate the sensitivity of *cartilage and synovial tissue*, Haller[60]*, introduced vitriol into the joint cavity of the pelvis of a live cat, at the point of linkage of the thigh-bone: another time, he introduced vitriol and antimony butter into the knee-joint. He pricked and burned the surface of these joints without the animal giving any sign of pain. Dörner[61], who under Autenrieth's** direction carried out 34 experiments on lesions of cartilage in living cats, investigating the walls of the nasal bone, and cartilage in the ear, larynx, ribs, and joints, makes no mention that the animals showed any sign of pain, except in Experiment 25, where 2 grains of lunar caustic were introduced into the knee-joint. This, of course, damaged the soft parts of the joint, and possibly some neighbouring regions.

Injury to the *arteries* appears to cause no pain of note. On no occasion did Haller[62] observe an animal crying out or giving other signs of pain if an artery were ligatured with a thread or wire at a site lacking nerves, and he refers to similar experiments carried out on humans

---

* A. von Haller (1708-1777), Professor of Medicine at Göttingen, where he made contributions to many branches of physiology including neurology.

** J.H.F. Autenrieth (1772-1835), Professor of Medicine at Tübingen.

by Bromfield* and Pouteau.** On the other hand when Bichat observed severe pain caused by the injection of irritating fluids such as ink, dilute acid, or wine into the arteries of living animals, this did not demonstrate sensitivity in the arteries: for these fluids penetrate through the pores of the vessel walls into neighbouring areas supplied with nerves.

Sinew areas, in the healthy state, are nearly or completely insensitive to external stimulation. Haller[63] quotes 39 authors investigating human beings, and 18 investigating animals, who carried out 200 experiments on the insensitivity of the sinews; he also names 25 authors working on humans, and 16 on animals, who made similar observations on the dura mater. [155] Observations on the insensitivity of the membranes covering the bones are also very numerous: it is known that during amputations, these membranes can be scraped off without producing pain[64]. In all these experiments the stimuli were partly mechanical (cutting, tearing, pinching), partly chemical (burning, or application of antimony butter, acids, alkalis, and other corrosive materials). Experiments involving the dura mater have even been carried out in the inflamed state. Bichat[65] claims to have observed that while the tendons, aponeuroses, and ligaments were insensitive to chemical and to most mechanical stimuli, severe pain could be evoked if they were violently stretched or twisted. In particular, it has been noted that ligaments pulled or strained during walking-tours can be extremely painful.

According to the experiments of Haller[66], Schobinger***, and Zimmermann, and confirmed by Bichat, connective tissue is insensitive when healthy. According to Bichat, it can, both in living animals and humans, be cut through, pulled in various directions, and stretched out by forced air without pain being caused, provided that the nerves running through it are not injured. Fatty cell tissue is also insensitive, generally speaking; but, in their experiments on human amputees, Duverny[67] and Monro found the *marrow* apparently to be sensitive, a finding confirmed in animals by Troja, Köhler, and Bichat. This sensitivity is however not a property of the bone-marrow, but doubtless of the tiny nerves - clearly demonstrated by Gros[68] - which penetrate the foramina nutritia along

* W. Bromfield (1712-1792), British surgeon and pathologist

** C. Pouteau (1724-1775), medical doctor of Lyons

*** S.S. Schobinger (1579-1652), J.G. Duverny (1648-1730), and M. Troja (1747-1828) were pioneers of medicine and surgery. G.H.E. Zimmermann and A. Monro were contemporary medical men of Weber's time. We have been unable to find information on Köhler and Gros.

with the blood-vessels and appear to run in part through the bone-marrow to the bone. [156]

The *bones*, in healthy condition, are insensitive. Bichat says that they may be sawn through, sectioned, rapped, or burned, without any observable pain being produced. Nevertheless, like many other regions insensitive while in a healthy state, they may become painful during disease as for instance in venereal or gouty diseases of the bone.

The umbilical cord also seems to be insensitive.

COMMON SENSIBILITY IN MAN AND IN ANIMALS, AND IN HEALTHY AND DISEASED SUBJECTS

Common sensibility is found in all animals, and even the embryo appears to change its position in the mother's body because of sensations of common sensibility. In patients, particularly hypochondrias and hysterics, stimuli which might be so weak as to be unnoticed by healthy persons, can cause vivid sensations of common sensibility. Many believe, unjustifiably, that the cause of these phenomena is an exaggerated sensitivity in the nerves. Rather, it would appear that various organic functions in such patients are even susceptible to slight stimuli due to various types of abnormality or affliction. Dysfunction then causes a further dysfunction, resulting in pain. In such debilitated persons, moderate exertion of the muscles rapidly brings on fatigue and pain: but this is not because the nerves are more excitable, but because the muscle substance is impaired and, after only very brief activity, undergoes changes which result in fatigue and pain.

FOOTNOTES

1. Bessel, *Astronomische Beobachtungen*. Section VIII. Königsberg 1835. Introduction. The difference between Bessel's results and other observers' could exceed 1 second. Struve, *Expedition chronométrique exécutée en 1843 entre Poulkova et Altona*. St. Petersburg, 1844, p. 29. According to these observations the difference did not exceed 3/10 second.

2. Galen, *De administratione anat.* lib. VIII. cap. 8. ed. Kühn, Vol II, p. 669, ed. Charter, Vol IV, p. 174, ed. Basil, Vol I, p. 187.

3. Valsalva, *De aure humana*, c. 5, section 8. See Tissot's works, III, section 183.

4. Felix Fontana, *Observations and experiments on the nature of animal bodies*, ed. v. Hebenstreit, Leipzig 1785, p. 138.

5. Joh. Müller, *Handbuch der Physiologie*. Vol I, p. 590, 4th edition 1843. E.H. Weber, On the effect of heating and cooling nerves on their ability to conduct, see *Berichte über die Verhandlungen der Königl. Sächs. Gesellschaft d. Wissenschaften*, Leipzig 1847, p. 175.

6. Magendie, *Journal de Physiologie exp. Paris*, 1825, Vol IV, p. 180 and 310-314.

7. Volta, in *Philos. Transact.* 1800, II, p. 405 sq.

8. Ritter, see J. Müller's *Handbuch der Physiologie*, Coblenz, 1837, II, p. 253.

9. Heydenreich, in *Frorieps Notizen*, 1848, Vol VIII, p. 34.

10. Joh. Müller, *Handbuch der Physiologie*, 3rd. ed., Vol I, p. 629, Vol II, p. 493.

11. Steinhäuser, *Experimenta nonnulla de sensibilitate et functione intestini crassi*. Leipzig, 1831, p. 19.

12. When I was carrying out the experiment already mentioned on p. [29], lying on my back and filling one.

nostril with ice-cold water through a tapered glass tube, this nostril was the first to fill up, then the upper part of the pharynx, and finally the other nostril, so that eventually the water filled both nostrils up to their openings. I could breathe and speak without the water running down into my throat. My nostrils and the adjacent cavities had a capacity of 16.6 cm^3 of water on one occasion, and 17.2 cm^3 on another. In a youth of 16 on whom I tried the same experiment, the nostrils had a capacity of 8.3 cm^3 of water on one occasion and 11.7 cm^3 on another. But I only felt the cold of the water at the entrances; it was much more faintly felt at the back of the nose, but then somewhat more clearly on the upper surface of the velum palatinum. I was not aware of the filling up of the other nostril with water and would not have noticed it had not my assistant drawn my attention to the fact that the water was rising in the other nostril. If the water was at $0^{\circ}C$, a peculiar pain was felt in the frontal cavity, but it had no resemblance to a feeling of cold. The young man noticed this pain most in the region of the lachrymal canal. Water which had been at $0^{\circ}C$ when the nostril was being filled, and was allowed to run out as soon as the nostril was full, had become so warm in the nostril in the meantime that if it was caught in a small container heated up to $20^{\circ}C$, it showed a temperature of $25^{\circ}C$.

13. Hildebrandt, *Handbuch der Anatomie des Menschen*, edited by E.H. Weber, Leipzig 1830, Vol I, pp. 275, 276, 281, 285, and 286.

14 Joh. Müller, in *Frorieps Notizen*, March, 1831, Vol 30, p. 113 and *Handbuch der Physiologie des Menschen*. Coblenz 1834, I, p. 665 ff.

15. Andral, *Précis d'anatomie pathol.*, Vol II, p. 281, transl. by Becker, II, 437, has collected 386 cases of vascular brain-damage: irrespective of consequent disease phenomena, there were amongst these cases:

202 cases in which bleeding started in the part of the hemipheres above the striate cortex and visual area, and also in these regions;

61 cases in the striate cortex only;

35 cases in the visual area only;

27 in the portion of the hemipheres above the centrum semiovale;

10 in the anterior lobes caudal to the striate cortex;

7 behind the thalamus n. opt. in the posterior brain lobes;

44 in the cerebellum pons, m. oblongata, spinalis, and pineal gland.

16. If it should ever be found that the muscles of one side of the body could be paralysed without any evidence of complete or partial loss of sensation, or that loss of sensation in the skin of one side of the body could be observed without any signs of paralysis or weakening of the muscles, we would be obliged to conclude that the sensory and motor nerves could be separated from each other at points beyond the occipital lobes and striate cortex, close to the convolutions of the brain. The separation is already known to occur in the occipital lobes and striate cortex.

17. To give a clearer illustration of how hemiplegia leads to a loss of sensation of touch in closely neighbouring regions of skin and quickly to a greater or lesser degree of 'common sensibility' as well, I shall give a short description of a special case.

G, 68 years old, playing-card maker, had suffered for about eight weeks from hemiplegia caused by exposure to draughts from open windows after having perspired heavily from hard work. The illness began when he suddenly found that he could no longer stand because his left foot had become weak. He is unable to move his left arm and foot: but the turning muscles of his head and trunk, the extensor muscles of his back, the bending muscles of his head and back, the masticatory muscles, the respiratory muscles, and the abdominal muscles, are unaffected. The patient cannot sit up in bed as the muscles from the leg up to the trunk are necessary for this: but when he is helped into a sitting position he can turn, bend and stretch. The pectoralis major and latissimus dorsi are therefore paralysed: the diaphragm, the intercostal muscles, and the sternocleido-mastoidei are not. The ability to experience touch or common sensibility is absent or weakened along the whole left arm and leg. Nevertheless there are isolated points in the instep, and in the calf between the tibia and fibula, where prodding with a fine-pointed knife can be felt, although this feeling is not one of pain as in the right foot, but of a touch or tickling eliciting a withdrawal on the part of the patient and occasionally a twitch of a single muscle in the leg in the region of the vastus ext. or the tibialis ant. Thus reflex movement can occur, showing that these muscles are not completely paralysed either. There are parts of the skin on which the patient has no feeling and others where contact with a hot spoon at a temperature of 51-56°R produces no pain, but does elicit a sense of contact. On the instep and in the anterior part of the calf he feels nothing if touched with a finger, and cannot distinguish warmth and cold: but he feels the touch of a hot spoon as a prick. However he is so doubtful about the location of the stimulation that he believes, for example, that he has been pricked in the calf when he was actually touched on the instep. On the left side of the abdomen, on the parts of the skin covering adjacent to the rectus abdominalis, he feels pricking not as pain, but as touch: he cannot tell whether he has been touched with the finger or pricked with a pin. He can distinguish high degrees of heat from cold. On the other side of the midline he is sensitive. On the left side of the back, if he is touched alternately with cold and hot spoons, he can feel the touch, but cannot distinguish between the hot and the cold, and is so unclear about the location of the contact that he thinks he has been touched in the small of the back when it is actually in the middle. On the side of the flank and on the glutaeus he cannot even feel the contact of a hot spoon. On the chest he can clearly distinguish hot and cold to the right of the midline, but not to the left; but he can feel needle-pricks as contacts along the line of the joints of the rib-cartilage. Peripherally close by, he does not feel them. On the neck he has no feeling of warmth or cold; but he does feel these on the face, near the mouth, on the jaws, and on the zygomatic bone. On the cheeks sensitive and insensitive regions are interspersed. The ear is without feeling: the burning of a hot spoon is not even felt as a touch. I did not further investigate feelings in the left arm.

18. Valentin, Über den Verlauf und die letzten Enden der Nerven. *Acta. Acad. Caesar. Leopold. Carol. Nat. cur.* Vol XVIII, p. 1, Tab VI, Fig 31 and 32.

19. E. Burdach, *Beitrag zur mikroskopischen Anatomie der Nerven.* Königsberg 1837, Tab II, Fig 3.

20. Valentin, loc. cit. Tab III, Fig 3.

21. Henle and Kölliker, *Über die Pacinischen Körperchen an den Nerven des Menschen und der Säugetiere*. Zürich 1844, 4.

22. G. Herbst, *Die Pacinischen Körperchen und ihre Bedeutung*, mit 16 lithogr. Taf. Göttingen 1848, 8, where the literature on this subject will also be found.

23. A. Vater, see Lehmann: *de consensu partium c.h. praeside A. Vater*. Vitembergae 1741. 4, recus. in Haller *Disp. anat. select.* II, pp. 970, 971.

24. Herbst, in d. *Gött. gel. Anzeigen*. Oct. 1848. St. 162, pp. 1670 ff.

25. Should likely be 0.013.

26. E.H. Weber, Panegyrin med. indicentis d. 13. mens. Nov. 1829. *Annotationes Anatomicae et physiologicae* Prolusio *VI*. p. 6. recus. sub Titulo: De pulsu, resorptione, auditu et tactu annotationes anatomicae et physiologicae auctore Ernesto Henrico Weber. Leipzig 1834, p. 149. When I found in 1829 that the accuracy of the sense of touch could be very exactly measured and compared at various parts of the skin, I asked my brother Eduard Weber, who then lived in Göttingen, to undertake a mutual investigation of the touch-sense with me and to come to Leipzig for some time for this purpose. This would have given the work a much greater degree of excellence. My brother was at first inclined to accept my offer, but unfortunately other scientific work prevented him from leaving Göttingen, and I was therefore obliged to forgo his assistance and undertake the work alone.

27. E.H. Weber, Panegyrin med. indicit d. 31 mens Maj. 1833 respecta Rud. Sachse diss. inaug. de scarlatina Prolusio *XIX*, Leipzig 1833 p. 7. Collected in the book: *De pulsu, resorptione, auditu et tactu annotationes anatomicae et physiologicae*. Leipzig 1834. 4. p. 149:

> "More nerve fibres terminate in fairly sensitive parts than in parts with a dull sense. When two impressions are made simultaneously on the same nerve fibre, only one common sensation arises, but when impressions are made on two fibres there is a double and differentiated sensation. The same nerve fibre undoubtedly brings sensitivity

to many parts of the skin, and that is how such a small number of nerves is able to sensitise the whole surface of the skin; for you will not discover spots that definitely lack sensation - not even by pricking the skin with an extremely sharp point. Indeed, the spots where adjacent impressions are not merged are arranged in the skin so that in the very sensitive areas there are many of them, while in the parts with poor sensitivity there are very few. Such spots are easily noticed: we are made aware of them by the prolonged use of the hands and other parts, and by varied contact. However, the more such spots are interposed on the skin between the two points of the compasses, the further apart the points seem to be separated."

28. Volkmann, *Neue Beiträge zur Physiologie des Gesichtsinnes*. Leipzig 1836, p. 50.

29. J. Müller, in *Archiv für Anatomie und Physiologie*, 1837, Jahresbericht CXXIX.

30. Hueck, Von den Grenzen des Sehvermögens, in *Müllers Archiv* 1840, p. 94 ff.

31. Grube, Über den Bau der Macula lutea des menschenlichen Auges, in *Müllers Archiv* 1840, pp. 39 and 40.

32. [Publisher's note in 1905 edition: The size of the letters and the distance between lines are approximately the same in the orginal edition as in the present.]

33. Joh. Müller, *Handbuch der Physiologie*, 4th ed. Vol I, Bk. III, Chapter 3, p. 594.

34. E.H. Weber, *De pulsu, resorptione, auditu et tactu*, page 41, reprinted from the programme: *Annotationes anatomicae et physiologicae*. Prolegomenon VI, d. XIII, Nov. 1829, p. 2.

35. Volkmann, in Wagner's *Handwörterbuch der Physiologie*. Art. Nervenphysiologie. Vol 10, p. 568.

36. Hueck, in *Müllers Archiv* 1840, pp. 86 and 87. The calculated sizes of the retinal image are so variable in Hueck's article that we must assume that there are several misprints or errors. For this reason I have not cited them but restricted myself to the angle of vision.

37. Valentin, *Lehrbuch der Physiologie*, Vol II, p. 428.

38. That the region of the retina with the most acute localization-sense is only 1/3 - 1/2 Paris line in diameter, and that within these regions only the centremost part has the finest acuity, is very important. This situation enables us to focus the eyes so that the extrapolated optical axes intersect at the visible object; for we involuntarily focus our eyes so that the object we wish to see forms an image at the most sensitive part of the retina. If therefore the sensitivity of the centremost point is reduced in one eye, with neighbouring regions of retina being more sensitive, a squint can develop. Prof. Ritterich of Leipzig has observed such cases. This arrangement however is also conducive to the indirect control of our awareness. What an effort it would be to control our attention directly if we were to see all the image-forming objects with equal sharpness. Moreover this arrangement facilitates the estimations we carry out when we move our eyes. On the other hand, however, we are prevented by this arrangement from perceiving many objects at a glance if the eye be kept still. Could the mathematician Dase's remarkable ability to estimate rapidly the number of objects placed side by side perhaps be due to the fact that the sensitive part of his retina is abnormally large compared with that of others? Such a gift might well have led him to start exercising his art and his calculation ability at an early age.

39. Comparisons of the various parts of the skin with respect to the sensitivity of their localization-sense is of particular interest because it is worth taking trouble to answer the question whether the ability to distinguish clearly degrees of pressure and temperature is distributed in the same way as is the localization-sense. I would particularly point out here that the first of these abilities is much more evenly distributed over the whole skin than is the localization-sense. This leads us to suspect that although the number of sensitive parts in equal areas of skin in various regions is not very variable, the number of elementary nerve-fibres giving these areas their sensitivity is highly variable in different parts of the skin.

40. My observations on the accuracy of the touch-sense in various parts of the skin were recently confirmed by Allen Thomson (in the *Edinburgh Med. and Surg. Journal*, No. 116). He says "I should note that I have carried out a considerable number of experiments, comparing the sensitivity of various parts of the skin, both on my own person, and on others, and I have obtained very nearly the same results. The

acuity of the touch-sense seems on the whole to vary more or less between different individuals, but with regard to the relative sensitivity of the various parts of the body I have found no particular discrepancies from the results published by Professor Weber." Valentin (*Lehrbuch der Physiologie des Menschen*, Brunswick 1844, Vol II, p. 565) who, with the collaboration of Theile, Gerber, Neuhaus and Bühlmann, has repeated my experiments, says "If the same regions of skin of several persons are compared, it will be found that the smallest critical distances can vary fourfold or more, in individual areas, particularly those with rather poor sensitivity. But the sensitivities of the various major regions relative to one another remain constant, or differ by very minor amounts: so that the tip of the tongue, for instance, in all the cases we have examined, has been 50 - 60 times more sensitive to touch than was the skin of the middle back."

41. See E.H. Weber, *De pulsu, resorptione, auditu et tactu. Annotationes anat. et physiol.* Leipzig 1834, pp. [135] and [137]. See also p. [48] above.

42. The discovery that elementary nerve-fibres do divide into several branches in the region of peripheral nerve endings does not contradict what I have said, namely that a single elementary nerve-fibre connects a specific area of the touch-organ with a specific area of the brain *via* a single nerve-fibre; if, as might be the case, these branches end at adjacent points on the skin (as they indeed appear to do).

43. Delezenne in *Recueil des travaux de la soc. des sc. de Lille* 1827 as extracted in *Bull. univ. des sc. nat.* XI, 275, and in Fechner's *Repertorium der Experimentalphysik*, Leipzig 1832, Vol. I, p. 341.

44. Feelings of common sensibility are, as I have shown earlier (p. [23]), sensations which we should consider to represent changes in our own sensory state: we should not relate them to objects. To the factors described in that section as rendering it impossible for us to relate our sensations to objects, must be added our knowledge that we cannot direct the attention at one and the same time to a large number of sensations, imagining them as being objects. But as numerous impressions are continually being made on our sense-organs, not all of which can be the focus of attention, we are in these circumstances only aware of an altered sensory state which such impressions cause and leave behind. Henle, however, seems to me to go too far when he supposes the

nature of common sensibility to reside exclusively herein: he says (*Allgemeine Anatomie*, Leipzig 1841, p. 728): "Common sensibility is the totality, the disparate chaos of sensations: which is fed to human awareness by all the sensitive parts of the body." Sensations which lead us to perceive changes in our sensory state can easily and directly - that is, without intentional associative forethought - be activated at will, and then seem either pleasant or unpleasant when we relate them to our desires. In addition, there are also very faint and hence unimportant feelings of common sensibility: feelings of moderate tension in our muscles are among these.

45. Ph. v. Walther in the *Journal der Chirurgie und Augenheilkunde:* die Lehre vom schwarzen Star und seiner Heilart. Vol XXX, p. 360.

46. Gerdy, see *Arch. gén. de méd.* Févr. 1847, p. 265 et sq.

47. Longet, see his discussion of this subject in *Archives générales de médecine*. Paris, 1847, p. 21.

48. Pirogoff, *Recherches pratiques et physiologiques sur l'éthérisation*. Petersburg, 1847, p. 21.

49. Beau, *Arch. gén. de méd.* Jan. 1848 in *Frorieps Not.* Apr. 1848, p. 136 .

50. Vieusseux, see in Mehlis' *commentatio de morbis hominis dextri et sinistri*. Göttingen 1818, 4, p.15.

51. For a look at the question of whether anaesthesia inhibits the ability to sense the degree to which our muscles are tensed, see the section on the investigation of common sensibility in the muscles (p. [149]).

52. Eduard Weber, *Handwörterbuch der Physiologie*, Art. Muskelbewegung. Vol III, Part 2, p. 3.

53. Bichat, *Allgemeine Anatomie*, translated by Pfaff. Leipzig 1803. Part II, p. 212.

54. Liebig, *Annalen der Chemie und Physik*. Vol 62, No. 3.

55. Bunzen, *Beitrag zu einer Künftigen Physiologie*, Copenhagen, 1805, p. 117. He electrically connected the nerves and muscles of the shank of a freshly-killed cow and, on closing the circuit, noted that a thermometer implanted in the muscle-tissue rose by several degrees.

56. Helmholtz, in *Müllers Archiv.*, 1848, No. 2.

57. Some cases of anaesthesia, which appeared in *Med. Chirurg. Transact.* Vol II, p. 217, and Vol III, p. 90, and in the *American Medical Repository* Vol IV, p. 225, will also be found in Nasse, *Zeitschrift fur psychische Arzte*, 1822, Part 2. An exceptionally interesting case not discussed there has been reported by A. Reid (see *Frorieps Notizen* 1829, Vol 24, p. 217). I shall recount it here. In 1802, a man named Walker fell from his horse. In 1812 he contracted a case of erysipelas in the leg. Both legs became numb and were insensitive to needle-pricks, and the sense of touch was lost. If he took a foot-bath Walker could not say whether the water was hot or cold until the leg had been immersed to above the middle of the upper thigh. He felt in this situation as if the foot were covered by a stocking or boot, or as if it had gone to sleep. Nevertheless he was completely capable of moving the feet. After 1815 he suffered a bruise of the metatarsal bone of the little toe, which caused the bone to become carious, and necessitated its removal. Walker asserted that he did not feel the slightest pain during this operation: it was much as if surgery were being carried out on a dead limb. The disease gradually spread: at the time Reid reported the case, sensitivity had been lost over nearly all the surface of the body, while motor functioning, though reduced, remained sufficiently intact for him to continue using his hands for cutting his food, writing, and holding the reins while riding: moreover he was able to walk for short distances without a stick. He remarked 'I feel nothing except with my mouth, that is, I am not able to say whether an object which I touch is hot or cold, rough or smooth.' He could not say whether he was holding the reins until he saw them. His feet felt as if they were stiff and heavy. His vision was poor, but in contrast, his hearing, taste and smell were good. No experiments were carried out to determine whether Walker could give the position of his limbs when he could not see them. But the fact that he could walk reveals that he still retained feelings of contraction in the muscles. This is also indicated by his assertion that his feet felt stiff and heavy. Sensations of touch and of common sensibility in the skin appear therefore to have vanished, but common sensibility in the muscles seems not to have been completely lost.

58. Longet, in *Comptes rend*, June, 1839, No. 23, p. 920.

59. Longet, *Anatomie und Physiologie des Nervensystems*, trans. by Hein. Vol. I, Leipzig 1847, p. 30.

60. Haller, De partibus c.h. sensibilibus et irritabilibus, *Commentar. soc. reg. Gotting.*, Vol. II, 1752.

61. Dörner, *De gravioribus quibusdam cartilaginum mutationibus*. Tübingen 1798, 8.

62. Haller, *De partium c.h. fabrica*, etc., Book II, Sect. 1, 12; also his *Second. mémoire sur les parties sensibles*, p. 217.

63. Haller, *Elementa Physiologiae*. Book XXVIII, in the Introduction.

64. Haller, in *Commentar. Gotting.* 1752, Vol II, p. 123 ff.; Opera min. I, p. 341, and *Castelli experimenta, quibus varias c.h. partes sensu carere constitit*. Göttingen 1753, sect. III.

65. Bichat, *Allgemeine Anatomie*, transl. by Pfaff. Vol II, Sect. I.

66. Haller, *Novi Comment. soc. reg. Gotting.* Vol III, p. 25.

67. Duverney, *Mém. de l'Ac. Roy. des sc. de Paris.* a. 1700, p. 199.

68. Gros, in *Comptes rendus*, Vol XXIII, No. 24, p. 1106, and in *Frorieps Notizen*, March 1847, p. 289.

WEBER'S INDEX TO DE TACTU

(This is found on pp. VI - VIII of the original text. All page numbers in this index refer to those of the original text.)

The sensations aroused in the touch organ are of two sorts: for we perceive (1) a certain pressure or traction (2) heat or cold; but in both cases the sensation seems to depend on some sort of pressure. p.[44 & 145]

On the correct perception of the place where we are touched

Two objects touching the skin simultaneously seem to be separated by a smaller distance the less sensitive the touch-sense is in various parts of the skin. p.[45 & 147] - A phenomenon worth noting from that observation is explained. p.[59 & 148] - Two objects simultaneously touching the longitudinal diameter of the arms or feet seem to us closer together than when touching the transverse diameter of those parts. p.[49 & 151] - The causes of this phenomenon. p.[151] - Two simultaneous impressions merge into one if the places touched are close enough together. We measure the sensitivity of the touch-sense by this distance, which is larger or smaller in different parts of the skin. A table containing experiments on this subject. p.[50] - A table showing the sensitivity of the touch-sense in different parts of the skin in lines of various length. For the parts named have a more acute touch-sense the shorter the lines appended to their names. p.[58]

If two contiguous and mobile parts are touched simultaneously, the parts touched are more clearly discriminated and seem farther apart than if one part is touched in two places simultaneously. Two parts of the skin touched simultaneously are discriminated more easily by us if they are located opposite each other or if they differ in structure or use. p.[61] - In the skin, spots possessing a more acute touch-sense frequently seem to be placed next to less acute spots. p.[62] - A strong impression obscures a less strong one. p.[62] - We discriminate impressions more easily in turn, if we are not touched at precisely the same moment of time. p.[63] - The parts in the extremities that excel with the most acute touch-sense are shown in a way approved earlier.

p. [63 & 157] - Which parts of the head excel in tactile sensitivity? p.[65 & 154] - Which parts of the trunk excel in tactile sensitivity? p.[67 & 155] - In many parts of the trunk two objects simultaneously touching the longitudinal line of the periphery of the trunk seem farther apart than when touching the transverse line; and this seems to depend on whether the course of the nerves is in a longitudinal or transverse direction. p.[68] - On the causes of tactile sensitivity. p.[75] - The touch-sense of the skin does not become more sensitive because we inspect part of the skin with the eyes, and in this way learn its arrangement and shape more accurately. p.[75] - Neither does sensitivity depend on the number and size of the papillae of the skin. p.[77] - Tactile sensitivity seems to depend instead on the number of nerves that terminate in a part of the skin. p.[77] - The touch-sense is much improved by voluntary movement. p.[77] - With closed eyes and a stationary hand we can scarcely (or not even scarcely) discriminate between very different objects placed on our finger-tips. p.[80] - We do not perceive the direction in which objects press on the skin, but we recognize very accurately the direction in which the hairs are pulled: this is done not only by the sense of touch, but by that sense joined together with the movement of the muscles when they resist the pull of the hairs. p.[78 & 158].

On the correct perception of the weight of objects.

The weight of objects is perceived in two ways: first by the touch-sense residing in the skin, and then by the sense belonging to the voluntary muscles, by which we recognize the degree of effort of the muscles in overcoming the weights and other obstacles. p.[81 & 159] - We do not perceive the value of the weights as clearly if the weights are placed on stationary and supported hands, as when they are weighed by our raised hands. p.[86, 134 & 159] - A certain weight on parts possessing a more acute touch-sense produces a sensation of stronger pressure. p.[94] - In those parts of the skin where we have the best tactile knowledge of the separation of the compass legs, a certain weight also produces a sensation of stronger pressure. p.[96 & 161] - A greater weight on a part of the skin possessing a blunter touch-sense arouses a similar sensation to a lesser weight on a part possessing a more acute touch-sense. p.[98] - All the organs of touch, even at their most sensitive, are very sluggish. p.[111] - For most people, a weight seems to make a greater impression when received by the left rather than by the right hand. p.[84].

On the correct perception of the temperature of objects

On certain errors to which we are liable in perceiving temperature. p.[113] - In those parts where we have the best tactile knowledge of the separation of the compass legs and of the value of weights, we also perceive the temperature of objects most clearly: it may therefore be supposed that the temperature and pressure senses reside in the same part of the touch organ. p.[163] - We perceive by the touch-sense an amount of heat either subtracted from or added to the skin, when it is at a certain temperature. p.[113 & 166] - If the palm of the hand has a temperature of 29 or $29\frac{1}{2}^{\circ}$R., water of 29°R. causes a sensation of heat, and water of 28°R. a sensation of cold. p.[169] - The sensations of cold and heat differ from each other in a different way than the sensations of light and dark. p.[169] - A lesser degree of heat received by a greater skin surface arouses in us a similar sensation to a greater degree of heat acting on a smaller surface; for impressions from single particles of water made on different spots of the skin merge together into one stronger impression. p.[113 & 165] - The back of the hand, when immersed in water, is more quickly affected by heat or cold than the palm. For its thinner cuticle does not resist heat or cold as strongly as the thicker cuticle of the palm; but after heat or cold has penetrated the cuticle the sensation in the palm is stronger, because of the greater number of nerves. p.[121] - We recognize a temperature difference more clearly when a larger skin surface is touched by warm objects than when a smaller surface comes into contact with the same object. p.[165] - For most people, a hot or cold object arouses a sensation of greater heat or cold in the left hand than in the right. The sensation of burning is a subjective sensation depending on too great an irritation of the brain produced by heat: it arises partly if the degree of heat is too great, and partly if the area of the body simultaneously affected by heat is too large. p.[170] - The less the degree of heat and the smaller the body surface exposed to the heat, the longer the temporal interval that elapses before a sensation of burning arises. p.[127].

On the comparison of the touch-sense with other senses.

The weights of two objects are more easily compared and more accurately discriminated in turn, if they are weighed by hand successively rather than simultaneously; the same is true of the discrimination of the temperature of objects, which are distinguished more clearly when perceived successively. The other senses are also subject to the same rule. p.[120 & 166] - The reasons is that a sensation merges with another simultaneous sensation and is confused with it, and so is perceived less clearly and is discriminated less well. Therefore we compare a past sensation with a present one more accurately than two present sensations with each other. p.[137] - The shorter the temporal interval between the perceptions to be compared, the more accurately they are compared: the more protracted the interval the more the comparison is degraded, till at last it cannot be performed at all. The difference between two weights is perceived no more easily in heavier than lighter weights, although the magnitude of the difference is greater in the former case than the latter. The same is also true of lines compared by sight and of tones compared by hearing; and so it follows that when noticing the difference between things to be compared, it is not the absolute difference that is perceived but the ratio of the difference to the magnitude of the things. p.[91, 163 & 172] - We do not notice a difference between weights *by touch* unless one is greater than the other by at least a *fifteenth* or a *thirtieth* part; we do not notice a difference between lines by sight unless one is longer by a *hundredth* part; finally, we do not notice a difference between tones by *hearing* unless one is sharper than the other by a *three hundred and twenty-second* part of a vibration. p.[142 & 172] - The senses are fatigued and tired by continuous work, so that we can no longer perceive differences which were formerly noticed very easily and clearly.*

---

* The reference is to p.[175], though no page number is given in the text.

INDEX OF NAMES

t - name appears in translator's footnote
w - name appears in Weber's footnote
italics - name appears in a reference list

SUBJECT INDEX

t - in translator's footnote
w - in Weber's footnote